Author

Debbie Snyder has been an educator and former homeschool mom of seven children for over forty years. Her teaching includes Bible, science, anatomy, health, speech, and debate in public schools, Christian schools, and homeschool co-ops. Her goal is to pass the torch to the next generation.

Illustrator and Adviser

Jeanette Summerville is a registered nurse, a former homeschool mom of eight children, an artist, and a dedicated missionary to the Deaf of Guatemala since 2010. With over 35 years of experience teaching and creating curriculum for both Sunday school and for the Deaf, she is passionate about making learning meaningful and accessible for all ages.

In Loving Memory
of
William Baker Snyder, Jr.
(1950-2021)

Dedicated

to
Snyder and Summerville Children (cousins)
David, John, Tom, Tim, Anna, Sarah, and Becca
Esther, Liz, Hannah, Nathan, Lydia, Josh, Luke, and Julia
And all of our grandchildren

Target Audience Tests

(Classes tested in-person and online)
Berryville Homeschool Academy: Berryville, VA (ages 7-9)
Classical Cottage School: Winchester, VA (ages 5-9)
Grandchildren: In-person and online (ages 3-8)
Guatemala Deaf Ministries: El Progreso, Guatemala (ages 5-14)

How to Use

Pick and choose activities according to abilities and interests.
As a unit study, this material is great for teaching
different ages at the same time.

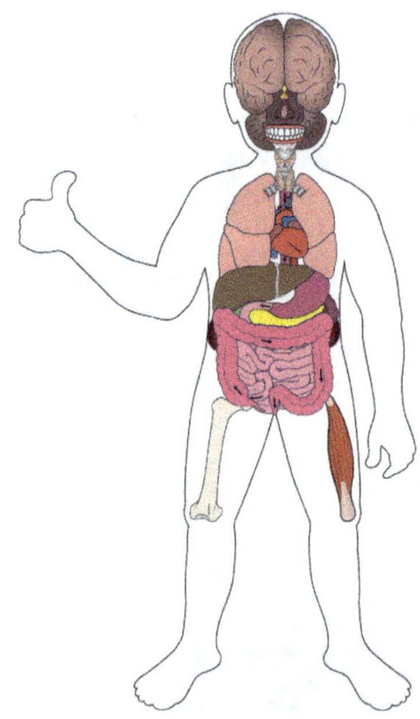

Copyright

Complete list of Images and Sources available in *My Awesome Body Teacher's Helps* (companion book).

Edited by Jeanette Summerville, Carrie Tucker, and Irina Snyder

exploringwithgrandma@gmail.com
exploringwithgrandma.com

My Awesome Body
Table of Contents

Twelve Body Systems
Overview

I will give thanks to You, because I am awesomely and wonderfully made;
Wonderful are Your works. And my soul knows it very well.
My frame was not hidden from You when I was made in secret,
And skillfully formed in the depths of the earth;
Your eyes have seen my formless substance;
And in Your book were written
All the days that were ordained for me,
When as yet there was not one of them.
How precious also are Your thoughts for me, God!
How vast is the sum of them!
(Psalm 139:14-17)

Vocabulary

Skeletal System: a system used to provide structure and give protection to body parts

Muscular System: a system used to give strength, posture, and movement by working with the skeletal system

Nervous System: a system used to carry messages between the brain and the body

Respiratory system: a system used to bring oxygen into the bloodstream and cells

Circulatory system: a system used to deliver nutrients and oxygen to all the cells in the body

Lymphatic System: a system used to fight infections with white blood cells; manages fluid levels

Urinary System: a system used to filter waste and remove excess fluid from the blood in the form of urine

Digestive System: a system used to break down food for energy, growth, and tissue repair

Immune System: a system used to fight diseases and infections

Integumentary System: a system used to protect from injury with skin, nails, and hair

Endocrine System: a system used to make the hormones that help cells talk to each other

Reproductive System: a system used to create hormones and new life

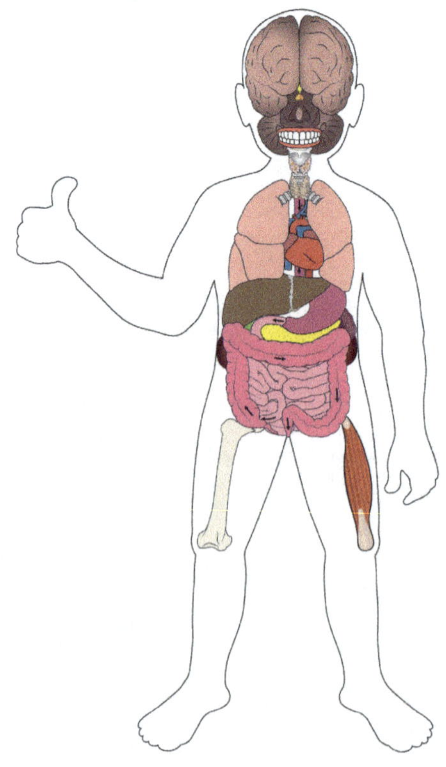

Body Systems Overview

For even as the body is one and yet has many members, and all the members of the body, though they are many, are one body, so also is Christ.
(1 Corinthians 12:12)

Purpose
To give an overview of twelve major body systems

Materials
Song: *Twelve Body Systems* (Hopscotch: YouTube)
Vocabulary (page 5)
Model or picture of a car

Procedure
Sing *Twelve Body Systems* by Hopscotch.
Discuss each system by using the vocabulary provided on page 5.
Emphasize the importance of all the systems working together in unity.

List the systems of a car using a picture or model (examples: brake, steering, and engine).
Highlight how car systems work together.
What will happen if just one system of a car does not work?
What will happen if just one system of a human body does not work?

Result
The concept of working together in unity will be highlighted.

Why?
Systems must work together with other systems to survive.

Matching Body Systems

Use Vocabulary (page 5) as needed.
The goal is to become familiar with each system.

1. Urinary System _____	A. Bones
2. Respiratory System _____	G. Babies and hormones
3. Nervous System _____	F. Muscles
4. Lymphatic System _____	L. Brain and neurons (nerves
5. Immune System _____	E. Fights diseases and infections
6. Digestive System _____	H. Skin, hair, and nails
7. Circulatory System _____	D. Filters liquid waste
8. Endocrine System _____	I. Food digestion
9. Integumentary System _____	K. Manages fluid levels
10. Muscular System _____	C. Lungs
11. Reproductive System _____	J. Heart and blood
12. Skeletal System _____	B. Glands and hormones

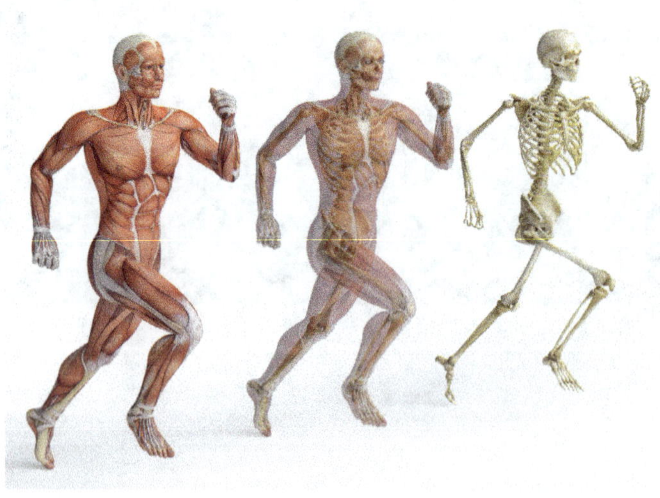

TWELVE BODY SYSTEMS
WORD SEARCH

Answers on page 136

A	I	N	T	E	G	U	M	E	N	T	A	R	Y
S	D	F	E	N	I	R	C	O	D	N	E	G	R
H	Y	R	E	S	P	I	R	A	T	O	R	Y	E
J	R	K	L	W	E	N	T	R	U	I	O	Z	P
X	O	R	C	V	B	A	E	N	M	W	E	R	R
E	T	Y	O	T	U	R	I	R	C	O	L	P	O
V	A	F	G	T	H	Y	J	I	V	A	K	L	D
I	L	D	E	S	A	A	T	Z	T	O	X	C	U
T	U	R	N	E	W	A	M	E	N	B	U	V	C
S	C	T	U	Y	H	U	L	I	O	P	L	S	T
E	R	Z	M	P	A	E	S	C	D	G	H	J	I
G	I	X	M	C	K	V	B	N	R	M	O	P	V
I	C	Y	I	S	W	A	S	C	G	I	V	B	E
D	L	R	R	A	L	U	C	S	U	M	C	N	M

URINARY	IMMUNE	INTEGUMENTARY
RESPIRATORY	DIGESTIVE	MUSCULAR
NERVOUS	CIRCULATORY	REPRODUCTIVE
LYMPHATIC	ENDOCRINE	SKELETAL

MAKE YOUR OWN
WORD SEARCH

Chapter 1
Skeletal System

Pleasant words are a honeycomb,
Sweet to the soul and healing to the bones.
(Proverbs 16:24)

Out of the ground the Lord God formed every beast of the field and every bird of the sky, and brought them to the man to see what he would call them; and whatever the man called a living creature, that was its name. The man gave names to all the cattle, and to the birds of the sky, and to every beast of the field, but for Adam there was not found a helper suitable for him.

So, the Lord God caused a deep sleep to fall upon the man, and he slept; then He took one of his ribs and closed up the flesh at that place. The Lord God fashioned into a woman the rib which He had taken from the man, and brought her to the man. The man said,

"This is now bone of my bones,
And flesh of my flesh;
She shall be called Woman,
Because she was taken out of Man."
(Genesis 2:19-23)

Vocabulary

Skeletal system: a system used to provide structure and give protection to body parts

Bone: the hard white tissue that makes up the skeleton

Skull: bones fused together to protect the brain, eyes, ears, and face

Vertebrate: protects neurons (nerves); used to help the body bend, twist, stretch, and stand

Bone marrow: the place where red blood cells, white blood cells, and platelets are created

Red blood cells: transports oxygen from the lungs to the cells and carbon dioxide from the cells to the lungs

White blood cells: cells created inside bone marrow to help fight disease

Platelets: cells that help wounds heal by forming blood clots to slow or stop bleeding

Joint: the place where two bones come together to move
a. hinge joint - moves back and forth like a door hinge (examples: elbow and knee)
b. ball and socket joint - moves like a joystick (examples: hip and shoulder)

Ligaments: bands connecting bones to other bones

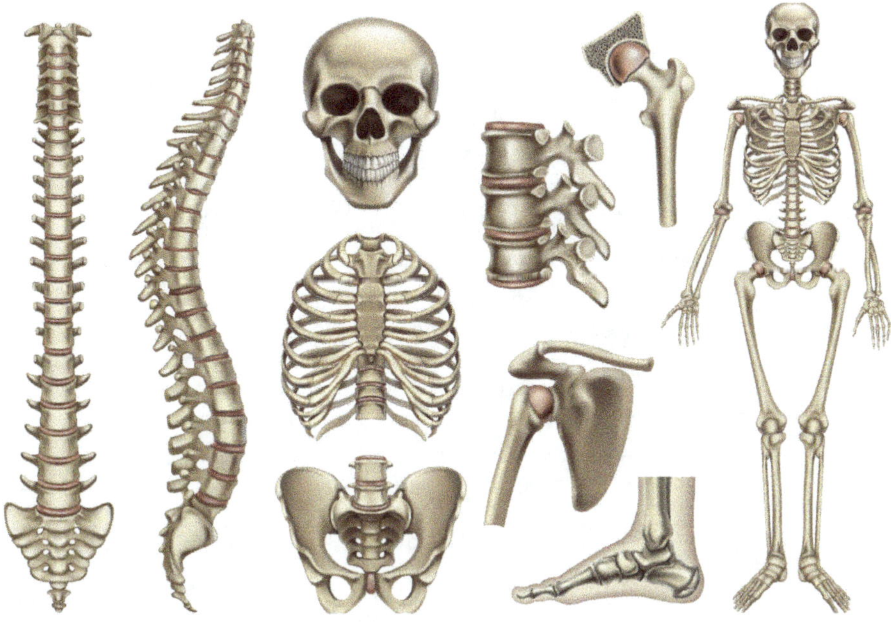

Skeletal System Song

Classic Tune: Dem Bones, Dem Bones, Dem Dry Bones

Verse 1

The cranium connected to the vertebrae.
The vertebrae connected to the rib bones.
The rib bones connected to the sternum.
Oh, hear the words of the health book.

Chorus

Dem bones, dem bones, dem dry bones,
Dem bones, dem bones, dem dry bones,
Dem bones, dem bones, dem dry bones,
Oh hear the words of the health book.

Verse 2

The sternum connected to the clavicle.
The clavicle connected to the scapula.
The scapula connected to the humerus.
Oh, hear the words of the health book.
Chorus

Verse 3

The humerus connected to the ulna and radius.
The ulna and radius connected to the the hand bones.
The hand bones connected to the phalanges.
Oh, hear the words of the health book.
Chorus

Verse 4

The pelvic bones connected to the femur.
The femur connected to the tibia and fibula.
The tibia and fibula connected to the foot and phalanges.
Oh, hear the words of the health book.
Chorus

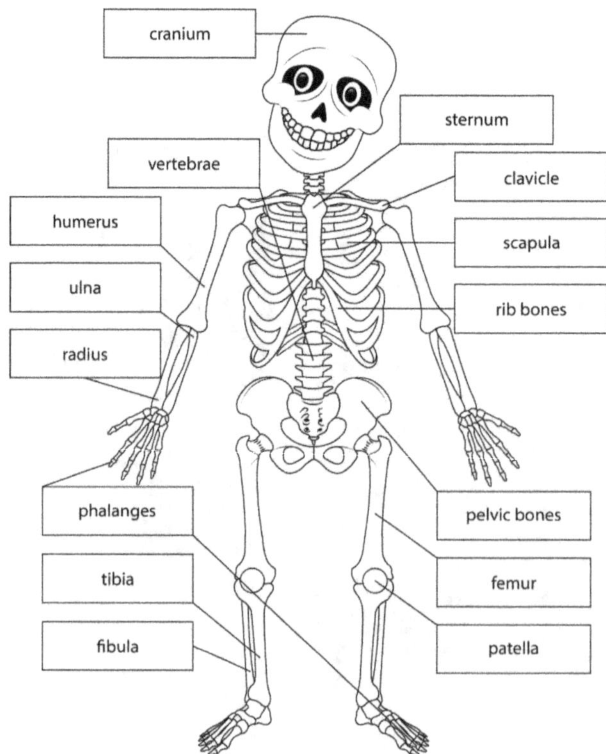

Skeletal System

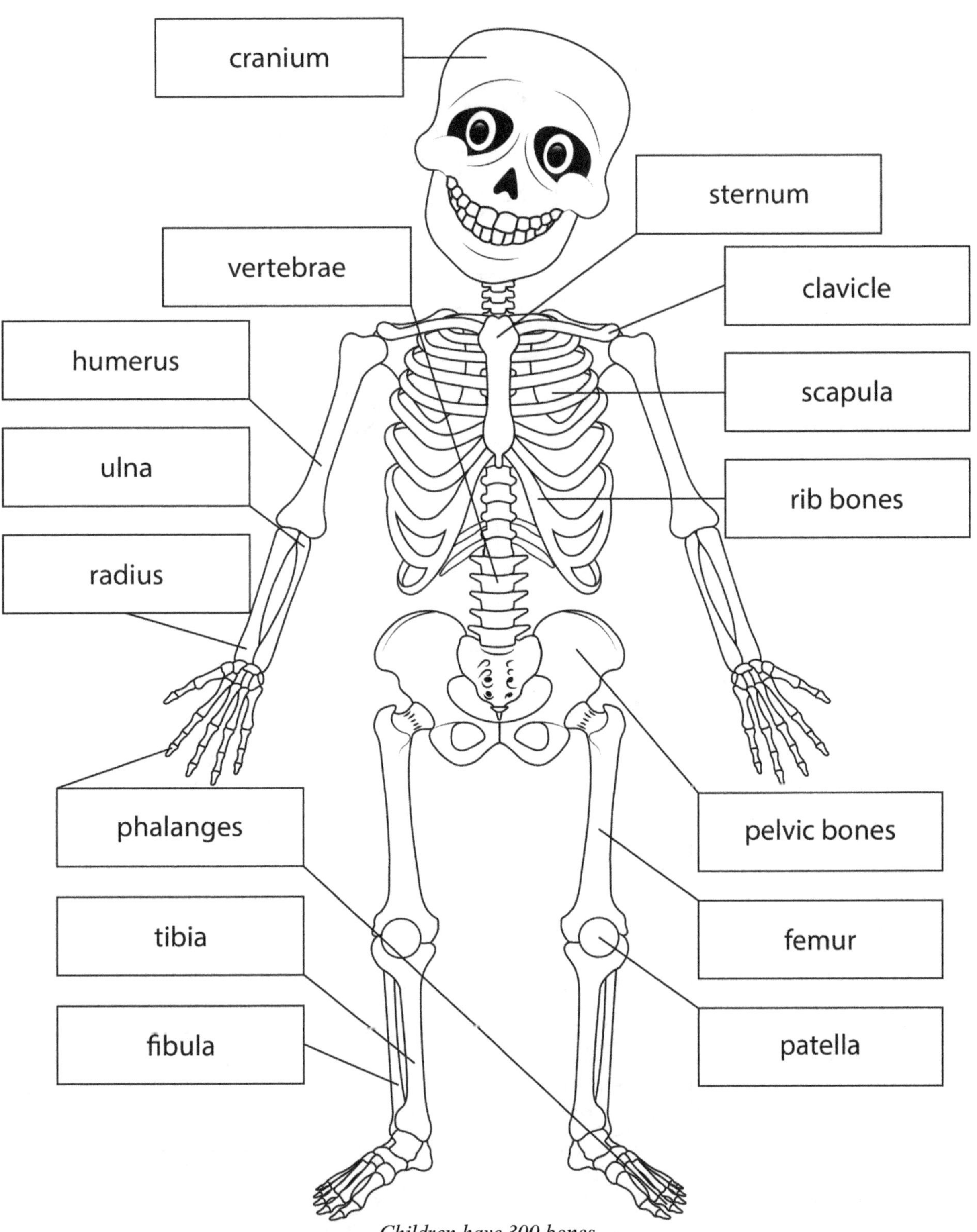

Children have 300 bones.
Adults have about 206 bones,
because some of their bones have fused together.

Activity

Label the parts of the skeleton using the diagram on page 12.

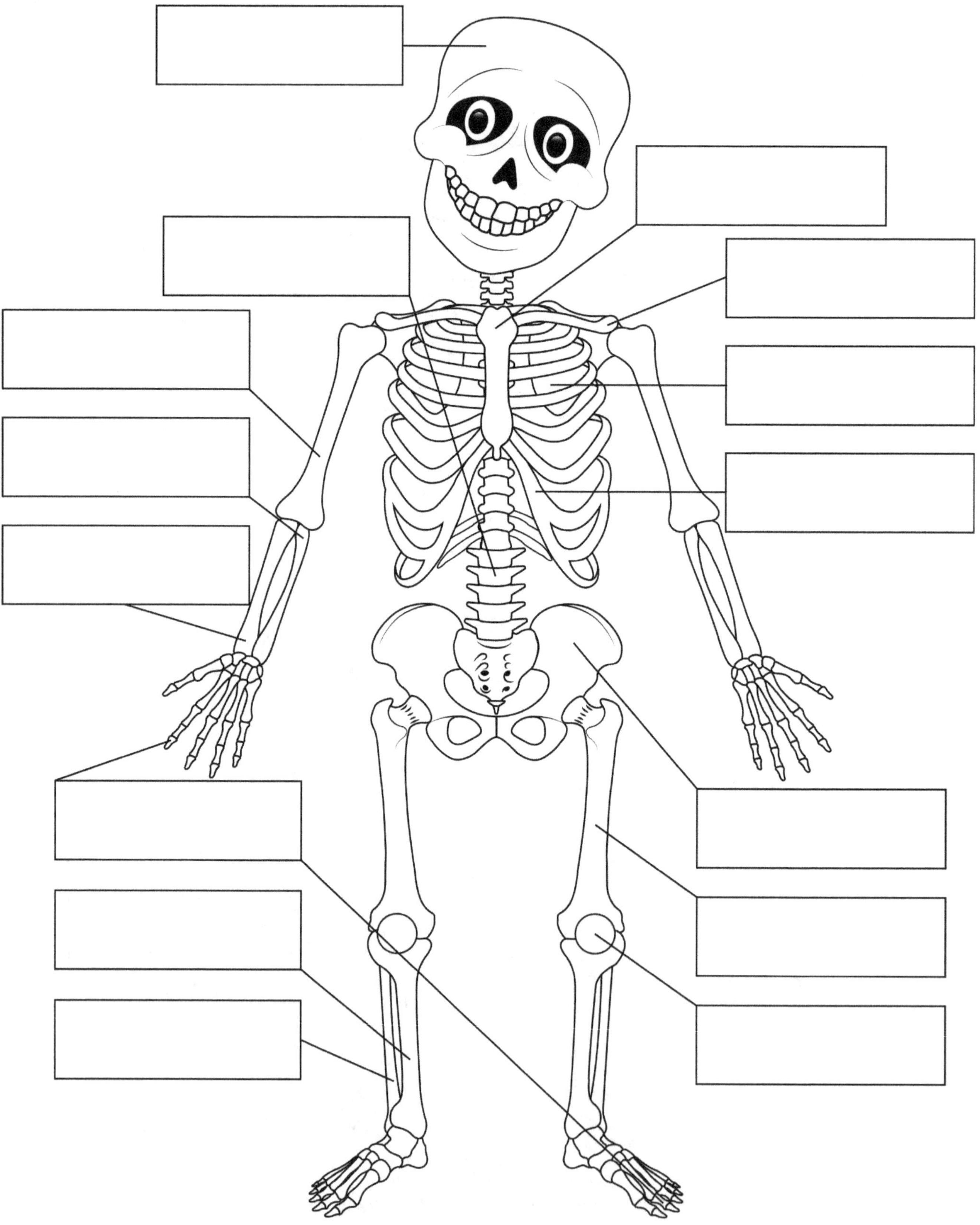

Vertebrates

How many vertebrates (bones) are in the neck of a giraffe?

How many vertebrates (bones) are in the neck of a person?

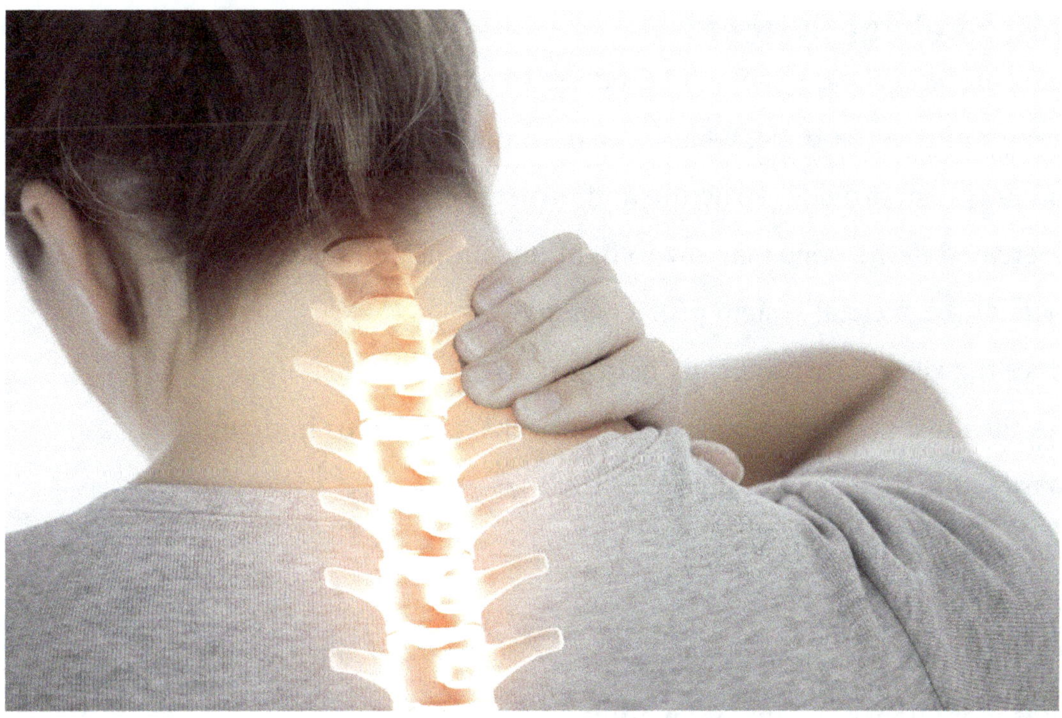

Answer: They both have seven vertebrates (bones) in their necks.

Skeletal System

Questions: Skeletal System

1. What part of the skeletal system did God use to create Eve?
2. According to Proverbs 16:24, what is sweet to the soul and healing to the bones?
3. What body system provides structure and protection?
4. What is created in the bone marrow to fight infections and diseases?
5. What part of the skeletal system protects nerves?
6. What does the vertebrate physically help you do?
7. Who has the most vertebrate bones in their necks, giraffes or humans?
8. What hard white tissue makes up a skeleton?
9. What is a joint?
10. What holds bones together?
11. What is the purpose of your skull?
12. Who has more bones, a baby or an adult?
13. What cells carry oxygen from the lungs to the cells via the bloodstream?
14. What cells help wounds heal by forming blood clots to slow or stop bleeding?

Skeletal System
Homework

Directions

These foods strengthen your skeletal system.
Circle the foods you ate or drank this week.

Yogurt

Milk

Cottage cheese

Eggs

Almonds

Dark green leafy vegetables

Sardines

Salmon

Fruits

Fortified cereal

Soy

Chapter 2
Muscular System

Behold now, Behemoth, which I made as well as you;
He eats grass like an ox.
Behold now, his strength in his loins
And his power in the muscles of his belly.
He bends his tail like a cedar;
The sinews of his thighs are knit together.
His bones are tubes of bronze;
His limbs are like bars of iron.
(Job 40:15-18)

Vocabulary

Muscular system: a system used to give strength, posture, and movement by working with the skeletal system

Muscles: the parts of the body that work with the skeletal system to contract and move; also provide stability

Voluntary muscles: the muscles that work by thinking (examples: biceps and triceps)

Involuntary muscles: the type of muscles that work without thinking (examples: digestive tract, lungs, and heart)

Skeletal Muscle: a type of voluntary muscle attached to bones allowing movement

Smooth Muscle: a type of involuntary muscles (examples: digestive tract and blood vessels)

Cardiac Muscle: the heart; an involuntary muscle

Tendons: used to connect muscles to bones

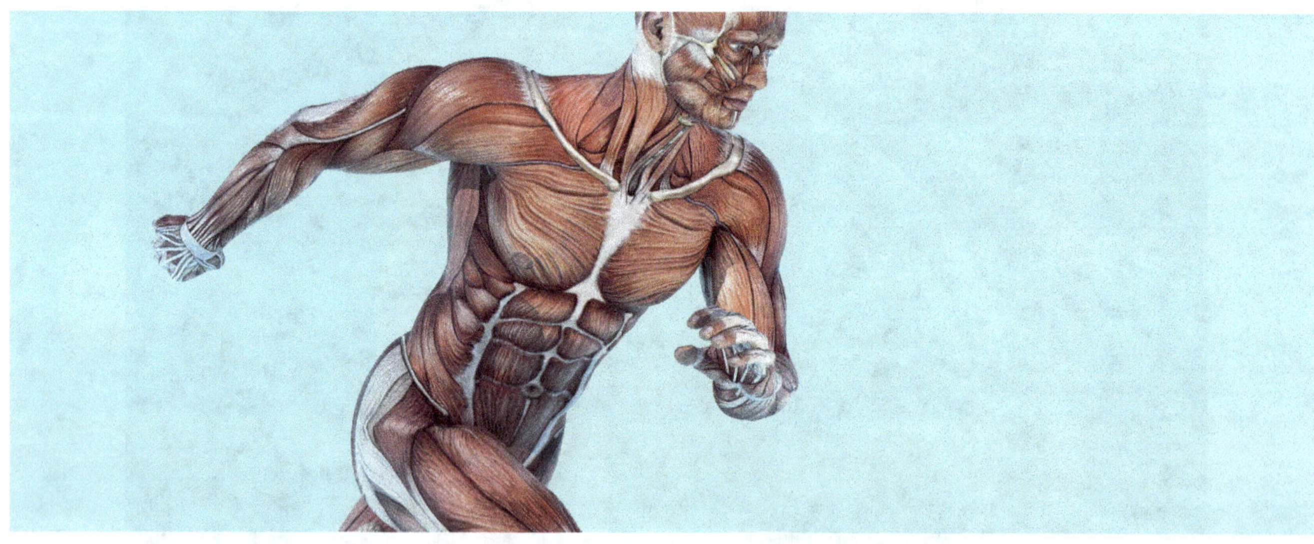

Muscle Trivia

There are about 600 muscles in the body.

Muscle building activities include walking, running, biking, swimming, and playing sports.

Biceps and Triceps

Biceps: "two" muscles on the top of the arms
Triceps: "three" muscles on the bottom of the arms

Quadriceps

Quadriceps: "four" muscles in the upper leg

Muscles

Purpose
To teach the difference between voluntary and involuntary muscles.
To learn the names and locations of biceps, triceps, and quadriceps.

Materials
Students

Procedure
Observe how blinking, breathing, and the beating of a heart are involuntary actions.
Observe how biceps and triceps can be moved by thinking.
Strengthen voluntary muscles with an exercise video. (See supplementary material)

Result
Students will be able to discern the difference between voluntary and involuntary muscles.
Students will be able to name the biceps, triceps, and quadriceps.

Why?
The muscles controlled by thinking are called voluntary muscles.
The muscles automatically working without thinking are called involuntary muscles.

MUSCLES
WORD SEARCH
Answers on page 137

A	V	S	M	O	O	T	H	B	C	D	E	F	G
S	Q	Y	S	K	E	L	E	T	A	L	G	D	A
T	R	S	P	O	M	N	L	B	I	C	E	P	S
U	V	T	W	X	Y	U	S	K	N	J	I	H	B
S	K	E	L	Z	A	P	S	B	V	T	C	D	C
W	C	M	V	B	E	M	L	C	O	R	P	A	Y
S	M	A	O	C	N	L	T	K	L	A	U	B	R
F	A	K	I	B	K	I	R	P	U	E	W	C	A
I	B	R	T	D	J	Y	Y	Z	N	H	X	D	T
S	T	M	O	I	R	R	W	E	T	M	A	E	N
I	U	K	G	U	L	A	N	R	A	M	D	F	U
O	S	E	N	O	B	W	C	H	R	S	P	S	L
P	X	C	V	B	N	A	M	V	Y	H	O	L	O
C	A	N	O	D	N	E	T	B	N	M	N	M	V

MUSCLE	SYSTEM	SKELETAL
HEART	SMOOTH	BICEPS
VOLUNTARY	CARDIAC	TRICEPS
TENDON	INVOLUNTARY	BONES

3 types of muscle tissue and cell

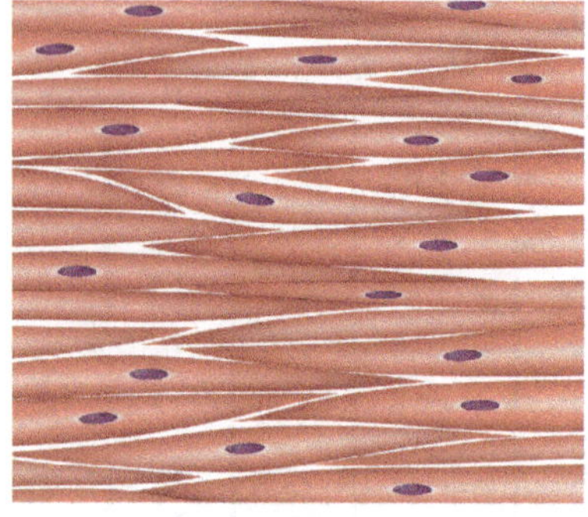

Smooth muscle

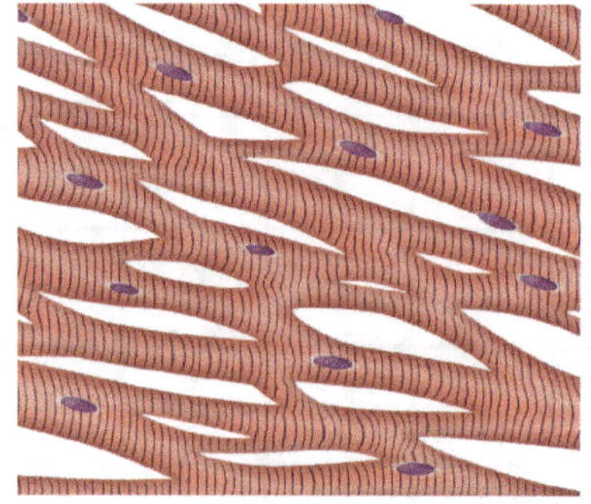

Cardiac muscle

Skeletal muscle

Questions: Muscular System

1. What are three types of muscles?
2. What is another name for the cardiac muscle?
3. What are two examples of smooth muscles?
4. What are two examples of voluntary muscles?
5. What are two examples of involuntary muscles?
6. What muscle pumps blood throughout the body?
7. About how many muscles are in a human body?
8. What is used to connect a muscle to a bone?
9. What two activities can help keep muscles strong?

Muscle Homework

Directions

Each of these foods strengthen your muscular system.
Circle the word if you eat or drink one of these foods this week.

Eggs
Milk
Cheese
Yogurt
Sardines
Salmon
Shrimp
Kale
Spinach
Collards
Broccoli
Swiss chard
Potatoes with skin
Broccoli
Boc choy
Oily fish
Poultry (chicken and turkey)
Lean red meat (beef and lamb)
Beans
Nuts and seeds

Eat Your Vegetables!

Chapter 3
Nervous System
Overview

Be anxious for nothing, but in everything by prayer and supplication with thanksgiving let your requests be made known to God. (Philippians 4:6)

Vocabulary

Nervous System: a system used to carry messages between the brain and the body

Brain: an organ inside the head that controls all the functions in the body

Spinal cord: the nerves, protected by the spine, used to send information to and from the brain

Neurons (nerves): the cells used to see, hear, smell, touch, and taste

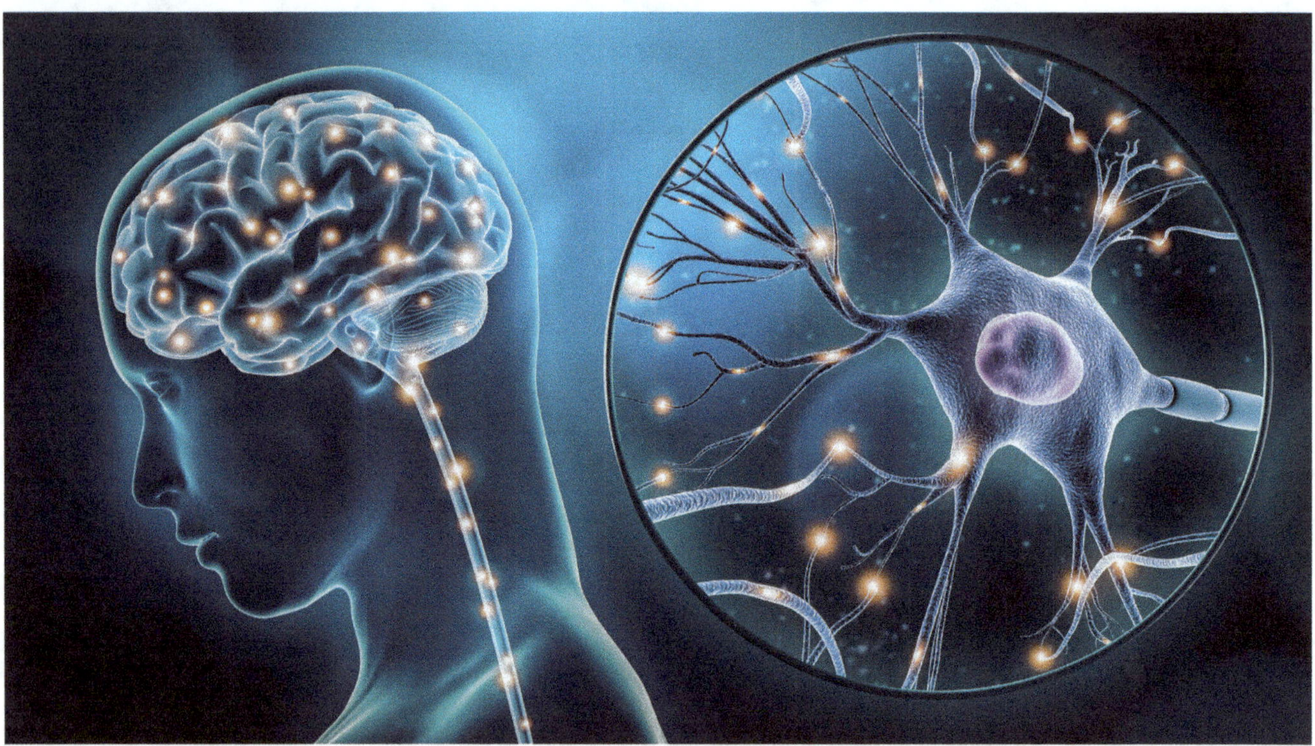

Five Senses
See, Hear, Smell, Touch, and Taste

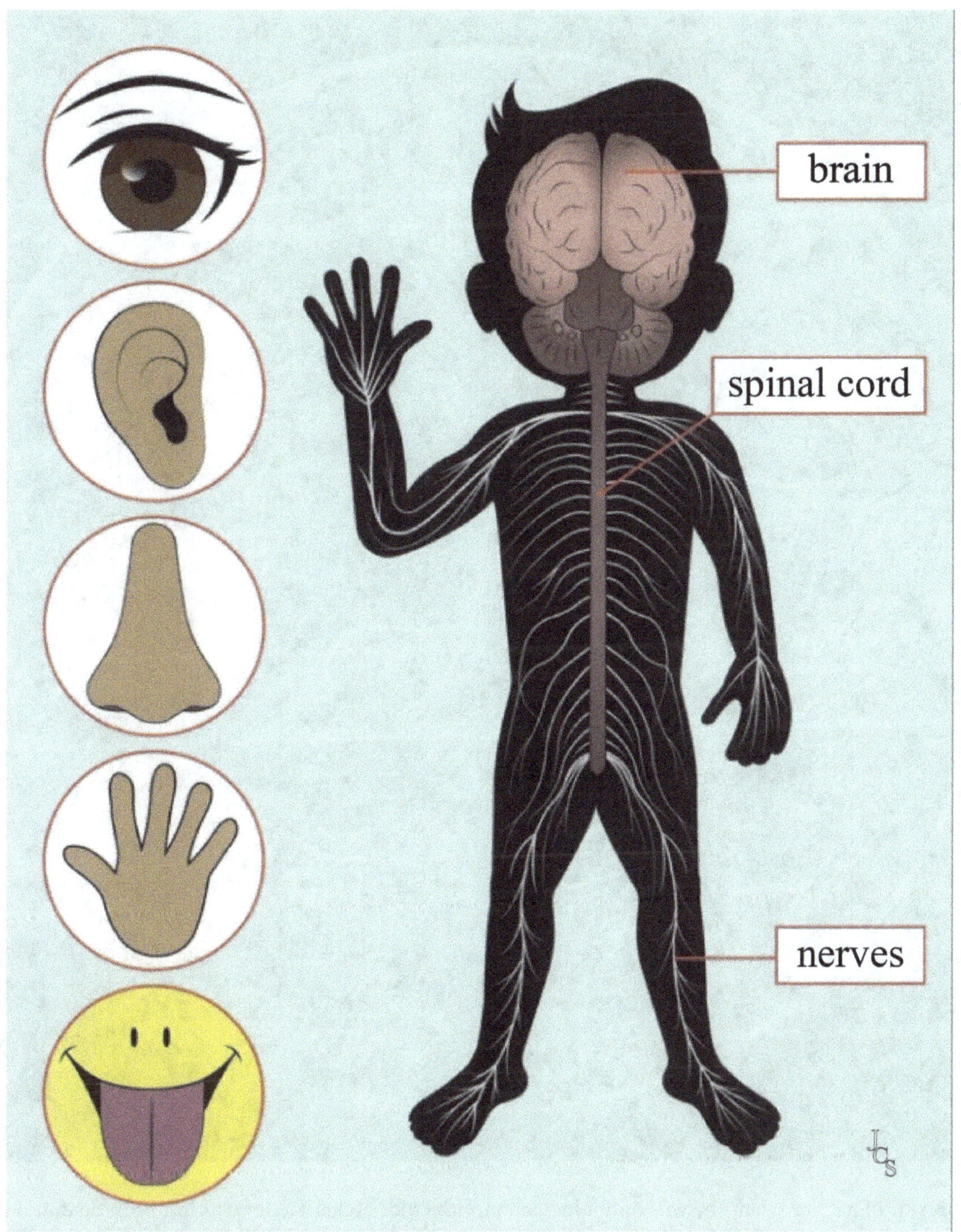

Questions: Nervous System

1. What body system contains the brain, spinal cord, and neurons (nerves)?
2. What organ controls all the functions in the body?
3. What kind of cells are used to see, hear, smell, touch and taste?
4. What is used to send information to and from the brain?
5. What are the five senses?

Helen Keller
(1880–1968)

As the result of a fever when she was only nineteen months old, Helen Keller was left both deaf and blind. She was thought to be unteachable and allowed to run wild. However, when Helen was seven years old, Anne Sullivan, herself partially blind, was hired to teach Helen. Anne's first goal was to get Helen to understand the connection between words and their meanings. A breakthrough came at a water pump when Helen understood that the finger signs Anne spelled in Helen's hand meant "water." It didn't take long before Helen learned the entire alphabet! Helen was not unable to learn. She just needed someone to "talk" to her in a way she could understand.

Helen Keller

Helen Keller was both blind and deaf; she could not speak words until she was grown. However, she was very smart and learned to read and write. Helen eventually learned to talk by feeling the vibrations of other people talking and became gifted at public speaking. She often used her words to speak up for others with disabilities.

Watch a version of The Miracle Worker. (YouTube)

Questions: Helen Keller

1. How old was Helen when she lost her sight and hearing?
2. How old was Helen when she was taught a way to communicate?
3. Who was Helen's first teacher?
4. Where did her breakthrough happen when she connected finger signs with language?

Nervous System
Homework

Directions

Each of these foods strengthen your nervous system.
Circle the word if you eat or drink one of these foods this week.

Leafy greens

Whole grains

Pumpkin seeds

Blueberries

Nuts

Avocados

Pomegranates

Sesame seeds

Olives

Beets

Tomatoes

Celery

You are what you eat!

Chapter 4
Sense of Sight
Nervous and Muscular Systems

Or how can you say to your brother, "Brother, let me take out the speck that is in your eye," when you yourself do not see the log that is in your own eye? You hypocrite, first take the log out of your own eye, and then you will see clearly to take out the speck that is in your brother's eye. (Luke 6:42)

Vocabulary

Eye: sensory organs that allow you to see

Cornea: the protective outer layer of the eyes

Pupil: the black opening in the center of an eye used to let in light

Retina: the part of an eye that senses light and sends images to the brain

Far sighted: distant objects are seen clearly, but nearby objects are blurry

Near sighted: nearby objects are seen clearly, but distant objects are blurry

Optical illusion: when eyes are tricked into seeing something that is not true

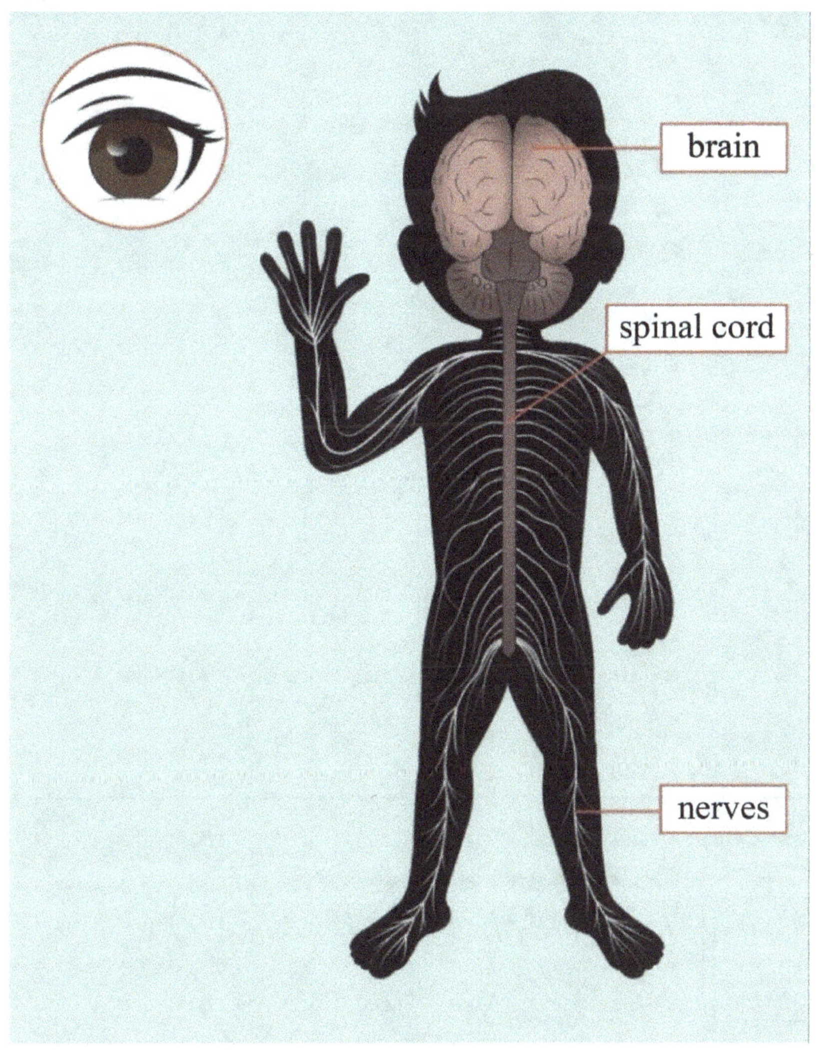

The eyes use both the nervous and the muscular systems to function.

Choose different parts to make a face.
Turn your paper over and try to draw the same face with your eyes closed.

Optical Illusions

Purpose
To discover the mystery of optical illusions

Materials
Optical illusions (see below)
Colored pencils, markers, or crayons

Procedure
Color the optical illusions.
Is it an old lady or a beautiful princess?

Result
It looks like an old lady at first,
but if you turn the picture upside down you can see a beautiful princess.

Why?
It is important to note that the world is not always what is seen at first glance.

Optical Illusions

The Eye

Questions: The Eye

1. What two systems does an eye use to see?
2. What is the clear, protective outer layer of an eye?
3. What is it called when our eyes are tricked into seeing something that is not there?
4. If a person can see distant objects clearly but nearby objects are blurry, is he/she far-sighted or near-sighted?
5. If a person can see nearby objects clearly but distant objects are blurry, is he/she far-sighted or near-sighted?
6. What is the black opening in the center of an eye called that lets in light?
7. What part of an eye senses light and sends images to the brain?

The Eye
Spiritual Application
Homework

Anyone who does not provide for their relatives,
and especially for their own household,
has denied the faith and is worse than an unbeliever.
(I Timothy 5:8)

Near-sighted: seeing the needs of "nearby" people clearly,
but the needs of "distant" people are ignored

Record blessings you provided this week for people in your own family.

1.

2.

3.

4.

5.

Love your neighbor as yourself.
(Matthew 22:39)

Far-sighted: seeing the needs of "distant" people clearly,
but the needs of "nearby" people are blurry

Record blessings you provided for people outside your own family.

1.

2.

3.

4.

5.

Chapter 5
Sense of Sound
Nervous System

So, faith comes from hearing, and hearing by the word of Christ.
(Romans 10:17)

Vocabulary

Ear: the part of the body used for hearing and balance

Outer ear: the part of the ear that can be seen

Ear canal: a passage made of bone and skin leading to the eardrum

Ear drum: the part of the middle ear that vibrates in response to sound waves

Cochlea: a structure in the inner ear that resembles a snail shell

Cilia: tiny hairs inside the cochlea that respond to different sounds

Auditory nerve: the nerve that carries an electrical signal to the brain from the ear

Vibrate: to move back and forth in order to create sound

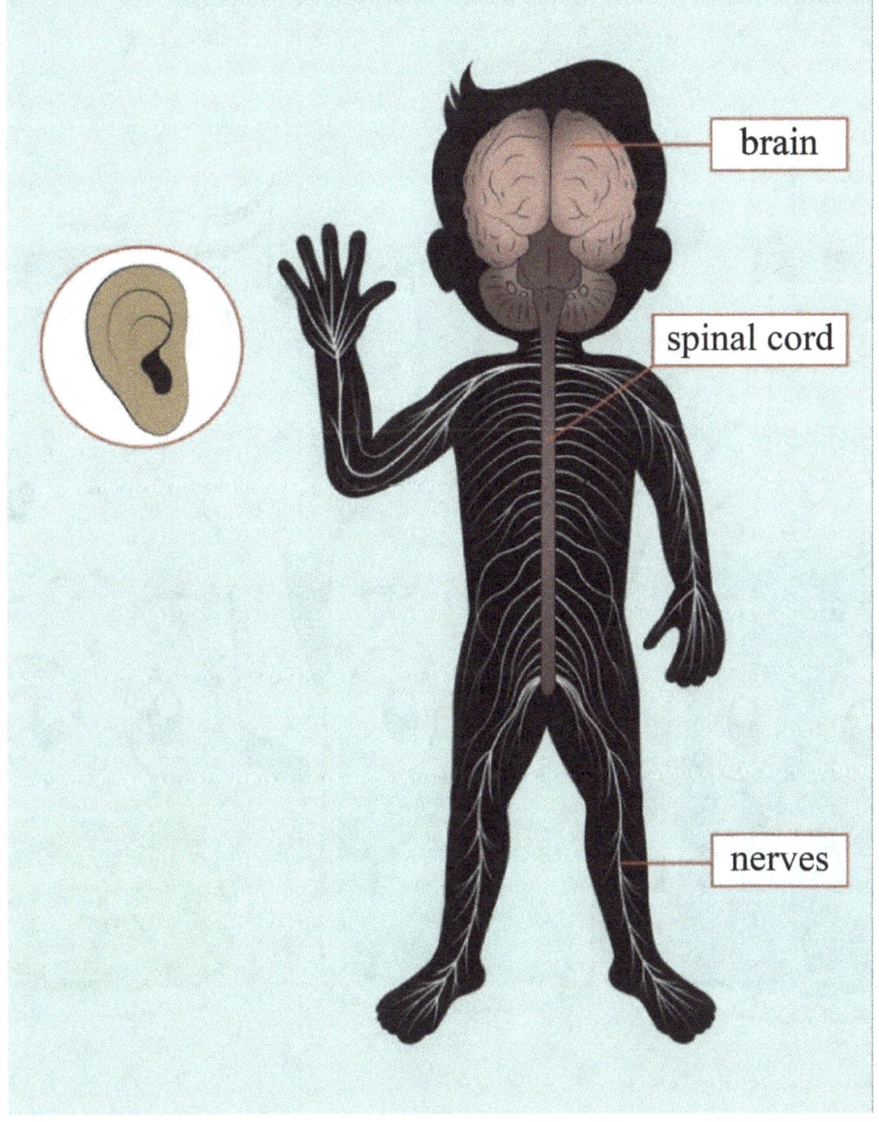

Music and the Ear
Activity

Purpose
To create music

Materials
Multiple jars
Food coloring (different colors)
Water
Spoon

Procedure
1. Place multiple jars on a table.
2. Fill the jars with different amounts of water.
4. Add food coloring to the water in each jar.
5. Using a spoon, gently tap the jars and listen to different pitches of "music."

Result
The differing levels of water will determine the pitch of the sounds.

Why?
The emptier the jars, the more quickly the sounds vibrate, resulting in higher pitch sounds. The fuller the jars, the slower the vibrations, which results in lower pitch sounds.

The Ear

Questions: The Ear

1. What part of the ear can be seen?
2. What passage leads to the eardrum?
3. What sounds are made by moving something back and forth?
4. What structure in the inner ear resembles a snail shell?
5. What vibrates in response to sound waves?
6. What (neuron) nerve carries sound signals from the ears to the brain?

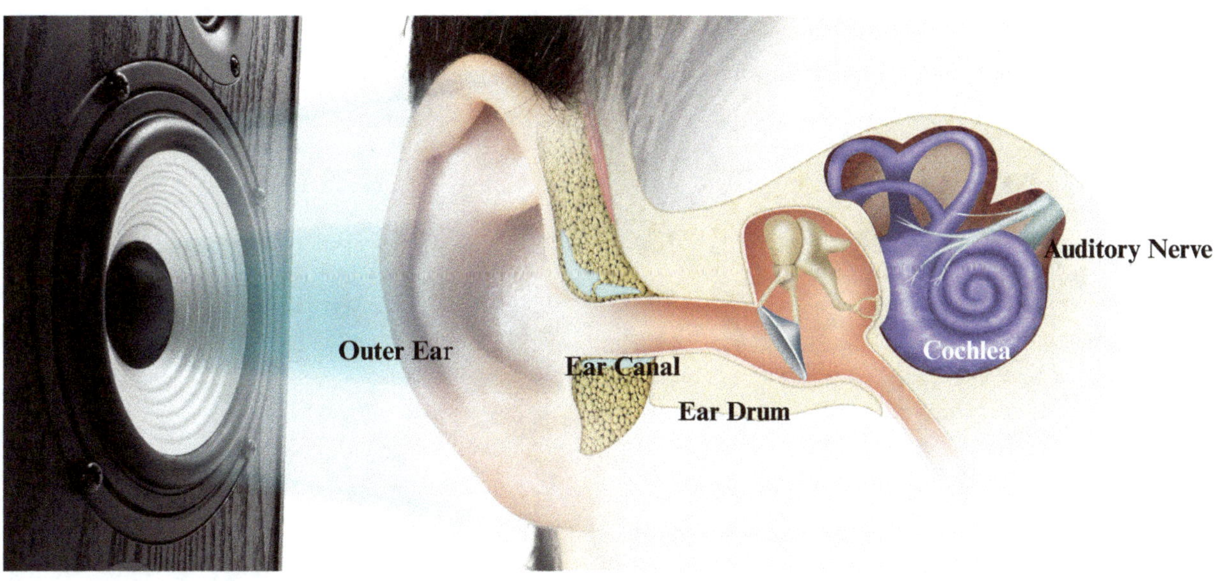

Sound
Homework

Draw or write ten different sounds you hear around your home.
(Examples: sirens, laughing, dishwasher)

1.

2.

3.

4.

5.

6.

7.

8.

9.

10.

Chapter 6
Sense of Smell
Nervous System

And he made the holy anointing oil and the pure, fragrant incense of spices, the work of a perfumer. (Exodus 37:29)

Vocabulary

Nose: the part of the body used for breathing and smell

Nostrils: two holes in the nose used for breathing

Nasal cavity: the space inside the nose responsible for the movement of air

Olfactory nerves: used to send messages to the brain to determine smells

Smell: the sense that detects odors using the nose

Cilia: tiny hairs that clean mucus from nasal cavities

Mucus: nasal fluid used to trap unwanted substances

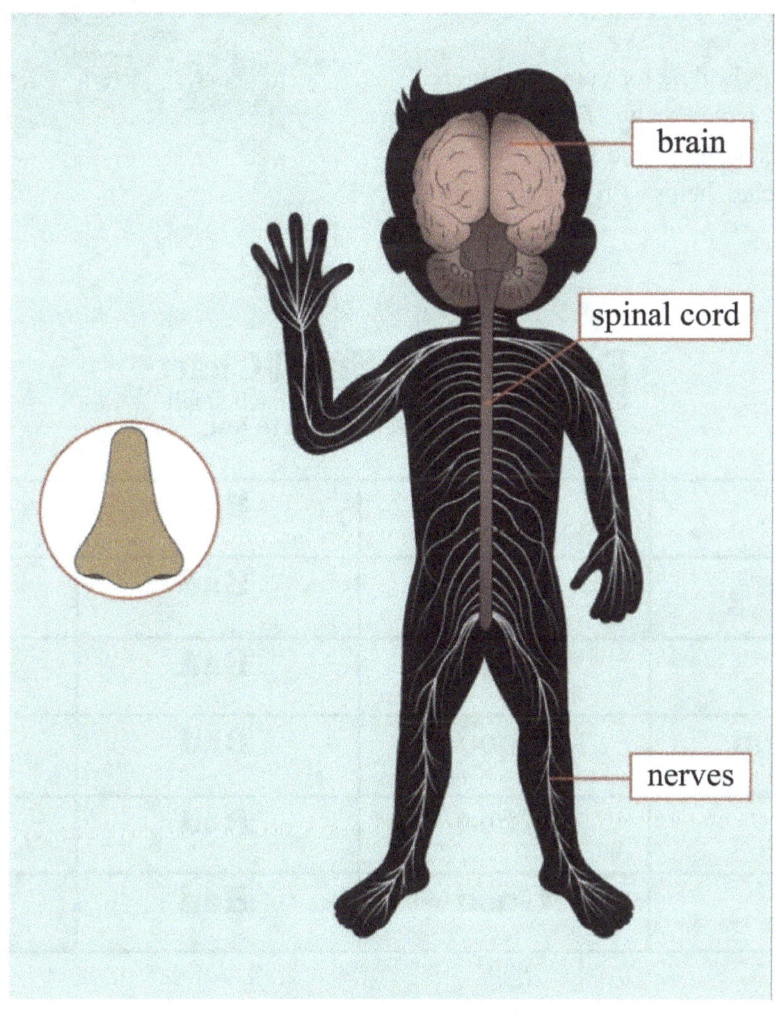

Trivia
There are over 10,000 possible smells.
Children smell better than adults.

Sense of Smell
Essential Oils

Purpose
To evaluate and rate fragrances

Materials
Various essential oils (examples: lavender, peppermint, clove, and lemon)
Smell Chart (see below)

Procedure
Explain different uses of essential oils.
Pass around different essential oils for smell evaluations.
Evaluate and circle: Good, Bad, or Neutral on the Smell Chart.

Result
The students will evaluate good, bad, and neutral fragrances.

Why?
One of the side effects of Covid-19 is the loss of smell and taste. Those who have had this illness are always grateful when their sense of smell and taste return.

Clove: flavoring, promotes healing for gums, and more

Lavender: helps sleep, promotes healing for burns, and more

Lemon: gives energy, kills bacteria, and more

Peppermint: eases headaches, helps with energy, and more

Essential Oils Smell Chart

Circle the word that describes each smell.
Choose other essential oils to test.

Clove	Good	Bad	Neutral
Lavender	Good	Bad	Neutral
Lemon	Good	Bad	Neutral
Peppermint	Good	Bad	Neutral
	Good	Bad	Neutral
	Good	Bad	Neutral

The African elephant has the best sense of smell on the planet.

Questions: Sense of Smell

1. The nose is part of what system?
2. What is the space inside the nose called?
3. What are the two holes used for breathing called?
4. What is used to send messages to the brain to determine smells?
5. Which sense is used to detect odors by the nose?
6. How many smells are possible?
7. Who can distinguish more smells, children or adults?
8. What animal has the best sense of smell?
9. What sickness can cause someone to lose the sense of smell and taste?

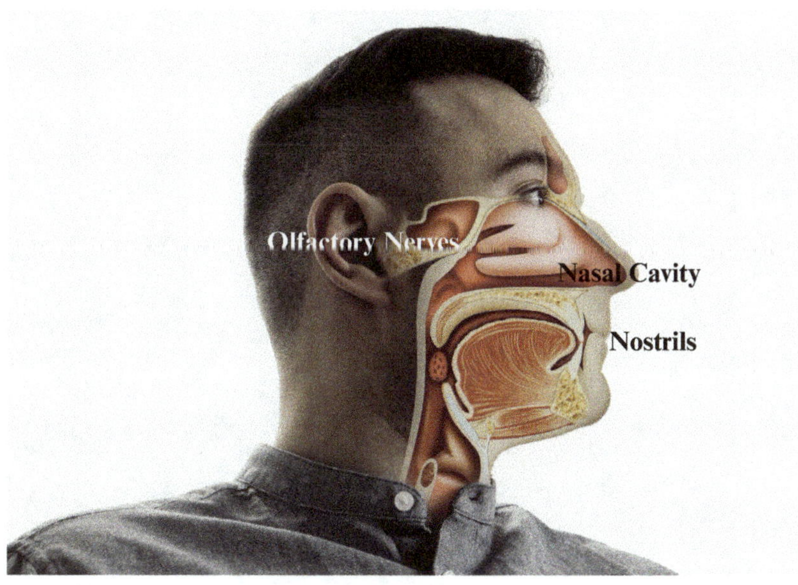

Sense of Smell
Homework

Notice ten different smells (examples: flowers, essential oils, coffee).
Write or draw them in the space below.

1.

2.

3.

4.

5.

6.

7.

8.

9.

10.

Chapter 7
Sense of Touch
Nervous System

And they were bringing children to Him so that He might touch them…
(Mark 10:13a)

Vocabulary

Neurons (nerves): the cells used to see, hear, smell, touch, and taste

Axon: the part of a nerve that takes information away from the cell body

Dendrite: the part of a nerve that brings electrical signals to the cell body

Cell body: the central part of a nerve cell

Nucleus: the control center of a cell

Myelin sheath: the fatty material that surrounds axons

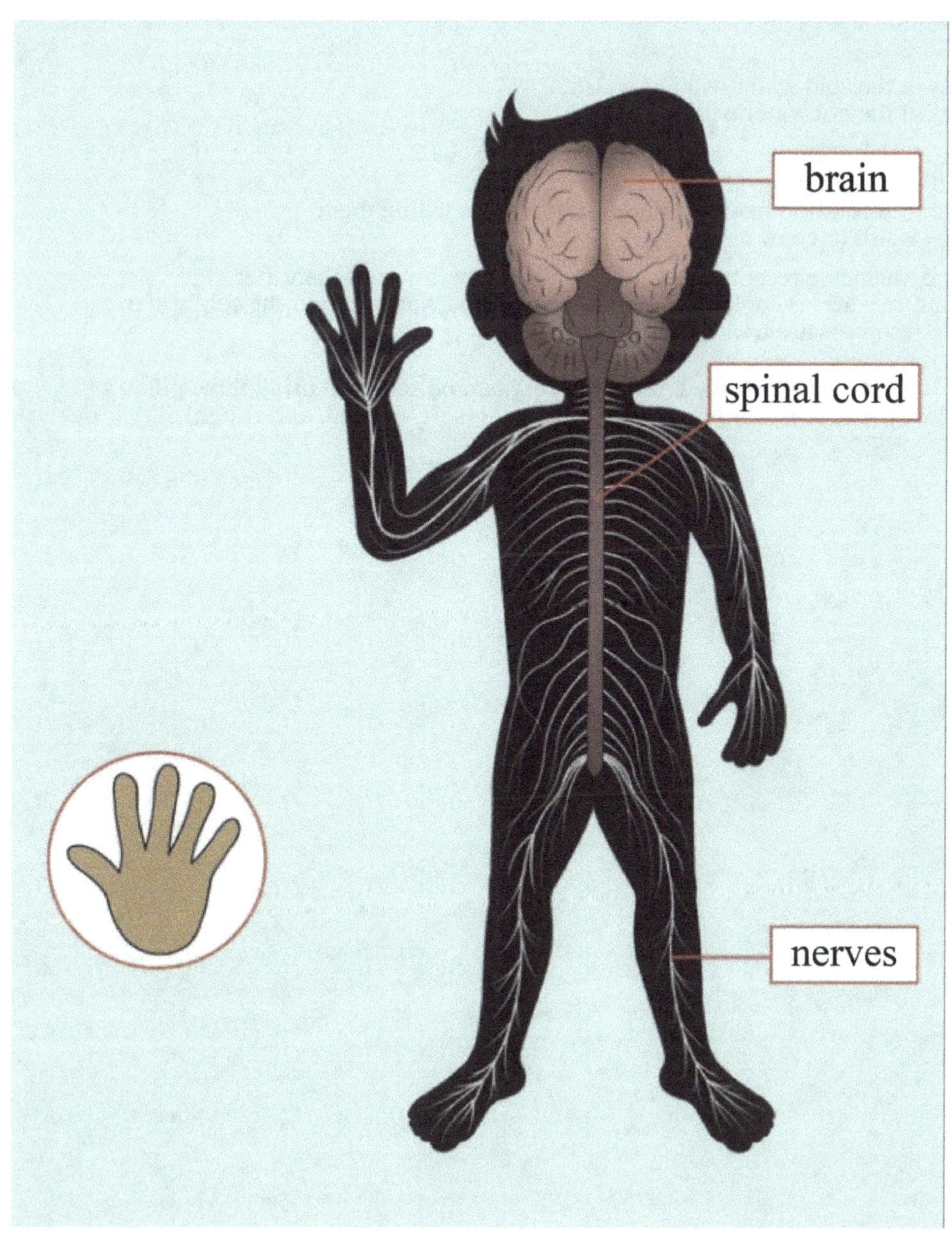

Touch
Activity

Purpose
To experience the sense of touch

Materials
Three plastic tubs
Water (hot, room temperature, and cold)
Ice

Procedure
Tub one: Fill with ice and cold water.
Tub two: Fill with water at room temperature.
Tub three: Fill with hot and room temperature water.
Note: The water should be as hot as possible without burning yourself.

Place one hand in cold water and one hand in hot water.
Count to thirty.
Immediately place both hands into the room temperature water.

Result
The hand that was in the cold water will feel warm.
The hand that was in the hot water will feel cold.

Why?
Tiny signals called nerve receptors send signals to the brain.
These receptors get information from temperature receptors telling them whether something is hot or cold.

Given enough time, the nerve receptors get used to the temperature of the water.
The room temperature water is cooler than the hot water and warmer than the cold water.
Thus, temperature receptors are tricked.

Note: The same kind of trickery can be achieved using three pillows by rubbing your hands on the textures (soft, neutral, and rough). Try it out.

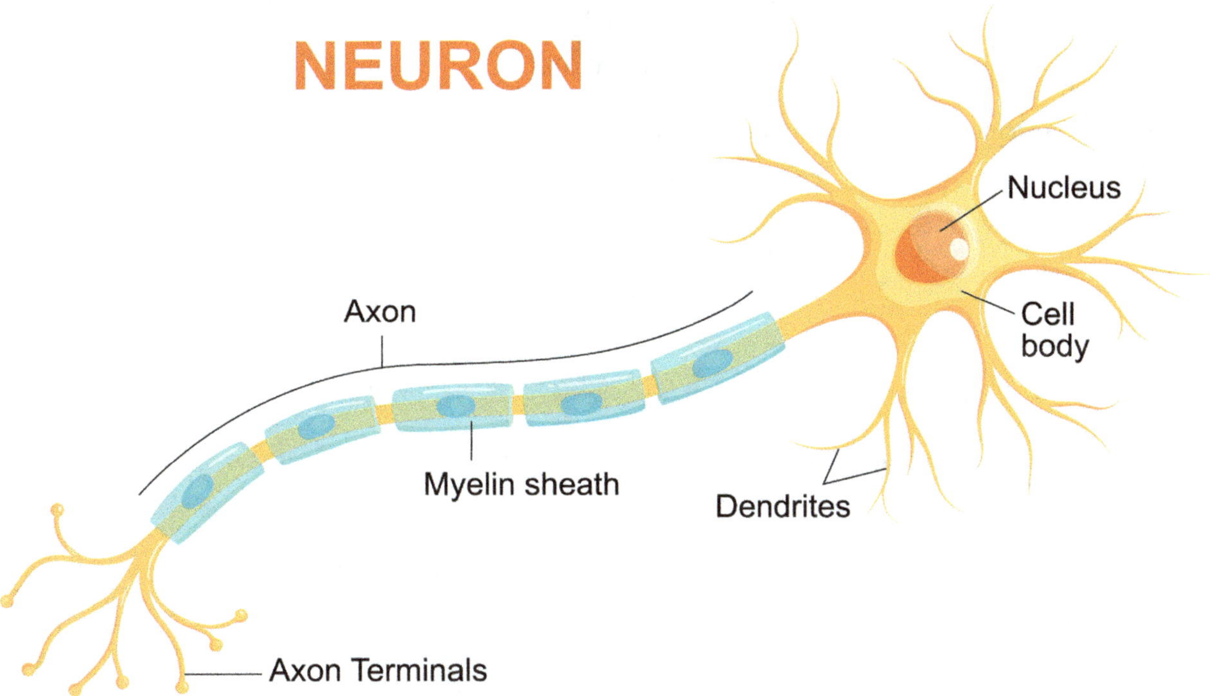

Questions: Sense of Touch

1. What system contains neurons (nerves)?
2. What part of a neuron (nerve) brings electrical signals to the cells?
3. What part of a neuron (nerve) takes information away from the cells?
4. What is the control center of a neuron (nerve) cell?
5. What fatty material surrounds the axons?

Using the picture above, draw and label a neuron (nerve) in the space below.

Sense of Touch
Homework

Draw or write the names of two things that are soft.

Draw or write the names of two things that are rough.

Draw or write the names of two things that are hard.

Draw or write the names of two things that are wet.

Draw or write the names of two things that are squishy.

Chapter 8
Sense of Taste
Nervous System

O taste and see that the Lord is good;
How blessed is the man who takes refuge in Him!
(Psalm 34:8)

Vocabulary

Tongue: the part of the mouth used for tasting, licking, swallowing, and speech
Taste buds: sensory organs on the tongue used for tasting
Bitter: harsh and disagreeable tastes (examples: unsweetened chocolate, coffee, and aspirin)
Salty: saline (examples: sea water, chips, and pretzels)
Savory: a spicy or salty taste without sweetness (examples: meat, cheese, and pizza)
Sour: having an acidic taste (examples: lemons, limes, and vinegar)
Sweet: food with a high sugar content (examples: cake, honey, and maple syrup)

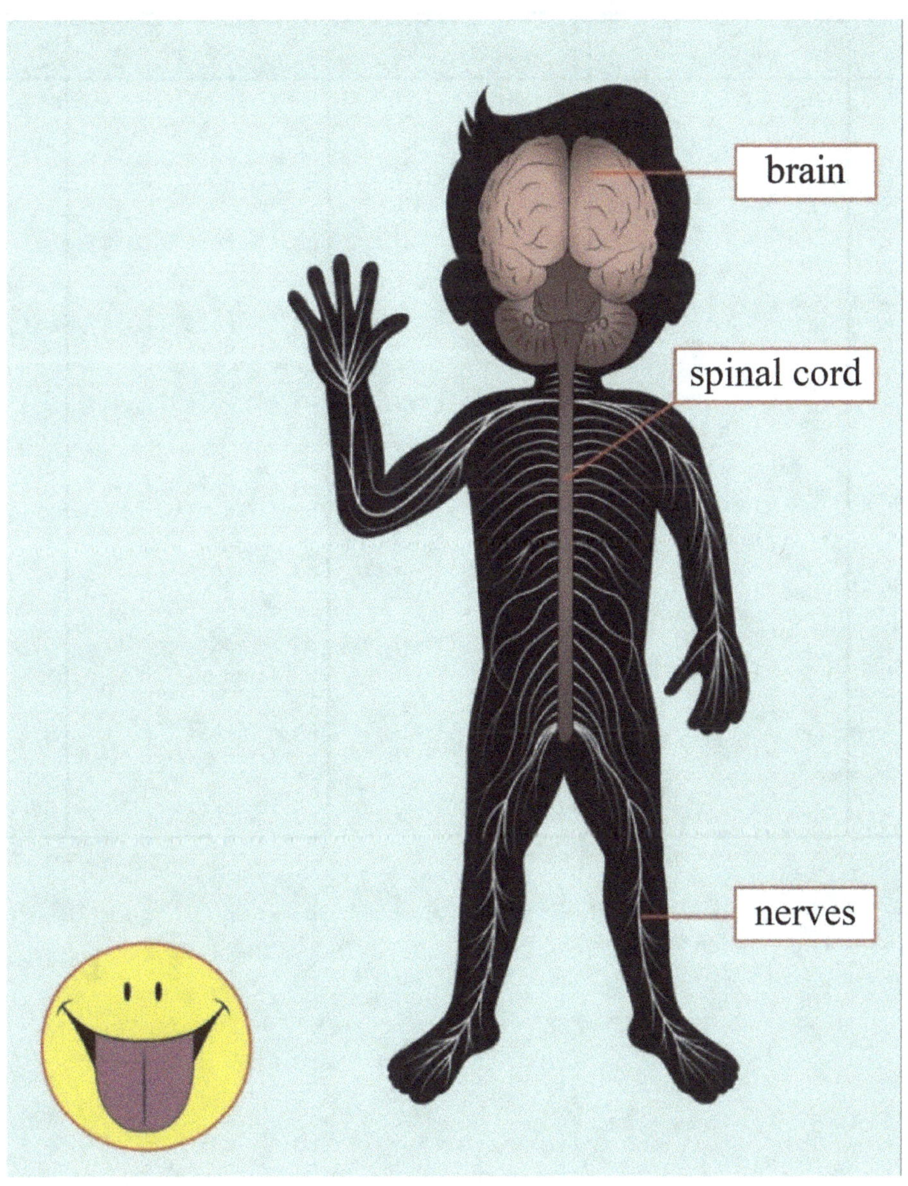

Taste

Purpose
To discover five categories of taste

Materials
Bitter: coffee and unsweetened chocolate
Sweet: honey and maple syrup
Salty: potato chips and salt water
Savory: pizza and cheese
Sour: lemon and vinegar

Procedure
Taste different types of food.
Write the names of the food under the proper categories on the chart below.

Result
Students will learn five categories of taste through experience.

Why?
The sense of taste helps us enjoy our food. Covid-19 causes people to lose their sense of taste and smell.

Sour	Sweet	Bitter	Salty	Savory

Taste

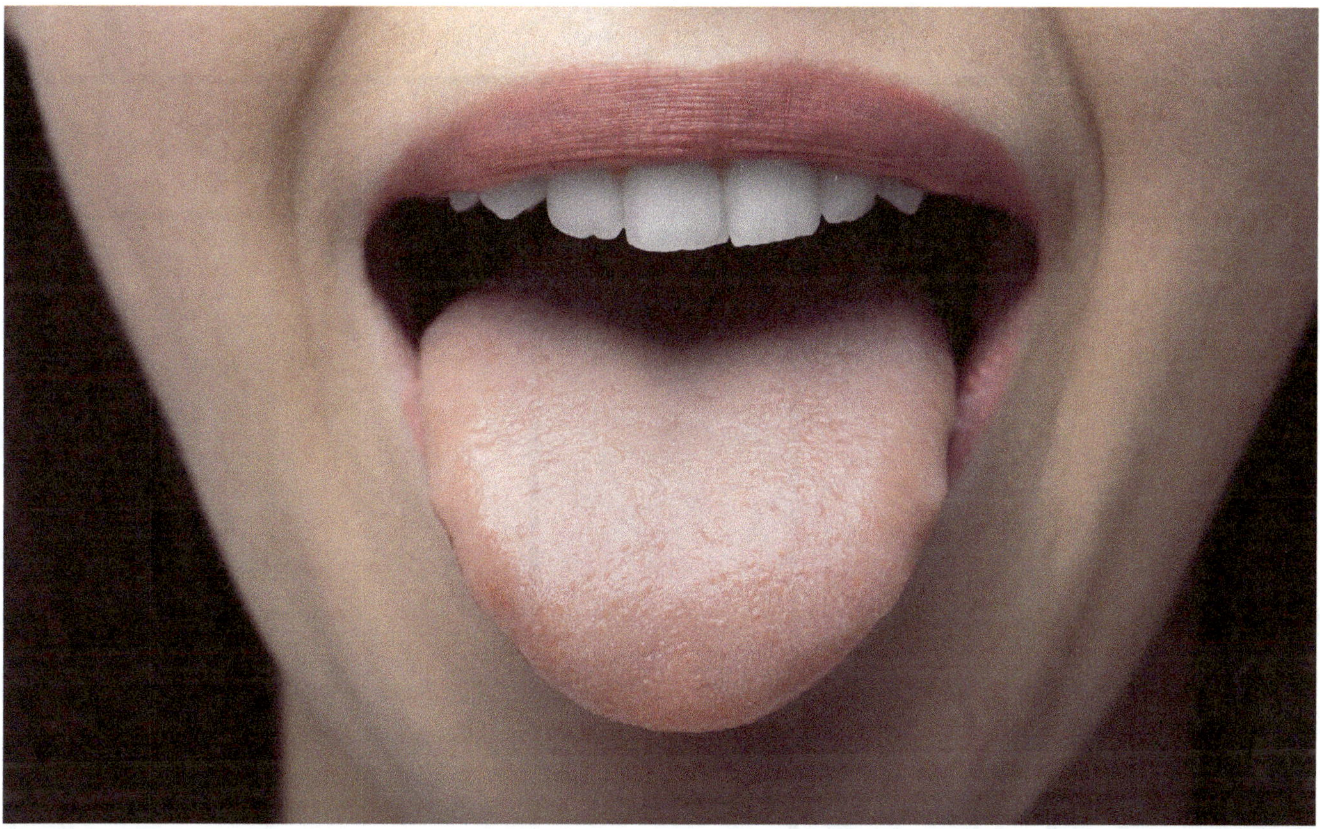

Questions: Sense of Taste

1. What are five different types of tastes?
2. What are the sensory organs found on the tip of the tongue?
3. How does unsweetened chocolate taste?
4. How do lemons taste?
5. How do pizzas taste?
6. How does honey taste?
7. How does sea water taste?
8. What disease causes some people to lose the sense of taste and smell?
9. What part of the mouth is used for tasting, licking, swallowing, and speech?

Sense of Taste
Homework

Chart some of the foods you eat this week into five taste categories.

Sour	Sweet	Bitter	Salty	Savory

Chapter 9
Jeopardy Review

Goal: To create **Questions** from **Answers** instead of **Answers** from **Questions**

On a white board, write the categories and point values provided below.

Divide the students into teams.
Use the **Answers** and **Questions** found on pages 51.
The teams will take turns until all the **Answers** and **Questions** are used.
Rotate students so everyone gets a turn.
Keep team scores on a white board.

Procedure: A student from each team will choose a category and a point amount.
The teacher will read the **Answer** connected to this choice .
This student will try to think of a **Question** that goes with the **Answer.**

If the student comes up with the correct **Question**,
he or she will receive full point value.
If the student cannot come up with the **Question,**
then ALL teams will concur with each other in separate parts of the room.

ALL the teams with the correct **Question** connected to the **Answer**, will get half the point value.
The **Question** can be written or whispered into the teacher's ear.

The game continues until all the categories and point values
have been chosen and the correct **Questions** have been provided.
The team with the most points wins.

Skeletal System	Muscular System	Nervous System
10	10	10
20	20	20
30	30	30
40	40	40
50	50	50

Jeopardy Questions

Skeletal System

10 Points
Answer: The part that a giraffe and man have in common
Question: What are seven neck (vertebrate) bones?

20 Points
Answer: The part of the skeletal system God used to create Eve
Question: What is a rib?

30 Points
Answer: The part of the skeletal system that protects the spinal cord
Question: What is a vertebrate?

40 Points
Answer: The number of bones in a child
Question: What are 300 bones?

50 Points
Answer: A type of blood cells that fight infections and diseases
Question: What are white blood cells?

Muscular System

10 Points
Answer: The muscle that pumps blood throughout the body
Question: What is a heart?

20 Points
Answer: Something used to connect muscles with bones
Question: What are tendons?

30 Points
Answer: The approximate number of muscles in a human body
Question: What are 600 muscles?

40 Points
Answer: A muscle used by thinking
Question: What is a voluntary muscle?

50 Points
Answer: A muscle used without thinking
Question: What is an involuntary muscle?

Nervous System

10 Points
Answer: The five senses
Question: What are sight, sound, smell, touch, and taste?

20 Points
Answer: The taste of coffee and unsweetened chocolate
Question: What is bitter?

30 Points
Answer: The part of the body used for tasting, licking, swallowing, and speech
Question: What is the tongue?

40 Points
Answer: Sensory organs on the tongue used for tasting
Question: What are tastebuds?

50 Points
Answer: A disease where people lose their sense of smell and taste
Question: What is Covid-19?

Chapter 10
Respiratory System

Then the Lord God formed man of dust from the ground,
and breathed into his nostrils the breath of life;
and man became a living being.
(Genesis 2:7)

Vocabulary

Respiratory system: a system used to bring oxygen into the bloodstream and cells

Trachea (windpipe): connects the throat to the lungs

Bronchial tubes: the airways used to let air in and out of the lungs

Lungs: the organs the body uses to breathe air

Diaphragm: a strong muscle under the lungs

Inhale: to breathe in air

Exhale: to breathe out air

Oxygen: an odorless gas used for energy, growth, and staying alive

Carbon dioxide: an odorless, colorless gas the body considers as waste

Oxygen comes into the body through the nose and mouth.
Then it travels down the trachea (windpipe), through the bronchial tubes, and then to the lungs.
Carbon dioxide is removed through the same passageways.

Activities

1. Lift your arm high into the air until it feels like it is burning.
(The burning is caused by oxygen leaving the arm.)

2. Go outside and run around, or run up and down steps.
(If you can do this for a minute without stopping, your lungs are in good shape.)

3. Have a contest and see who can hold their breath the longest.
The winner has the strongest lungs.

4. Lungs are like billows. As they expand, they inhale air for oxygen. As they compress, they exhale carbon dioxide.

Interesting Trivia: Sometimes you may swallow and cough because something seems to have gone down the "wrong pipe." Our throat has two "pipes": the trachea (windpipe), connecting the mouth to the lungs, and the esophagus, connecting the mouth to the stomach.

Make a Lung
Activity

Purpose
To demonstrate how your lungs work
(See supplementary material for similar demonstration.)

Materials
White cardboard
2 copies of the lungs and face; page 54)
Black marker
2 balloons
Two plastic straws
Tape
Glue

Procedure
Print 2 of the lungs and face
Handout 1: Color the lungs (pink) and the trachea (orange)
Trace the printed lines with a black marker; put aside
Handout 2: Color the face only; cut it out.

Attach a balloon at the bottom of each straw with tape to prevent the air from leaking.
Handout 1: Place a balloon/straw on top of each lung and tape it down at the top of the balloon.
The straws should meet at the top. Tape them down there too.

Tape or glue the colored face on top of the straws where the face should go.
Glue completed project onto a piece of cardboard.

Demonstrate inhaling and exhaling by breathing through the straws at the top.
Label the picture LUNGS.

Result
Inhaling and exhaling will be demonstrated.

Why?
Lungs bring oxygen to all parts of the body.
No one can live without lungs.

Make a Lung

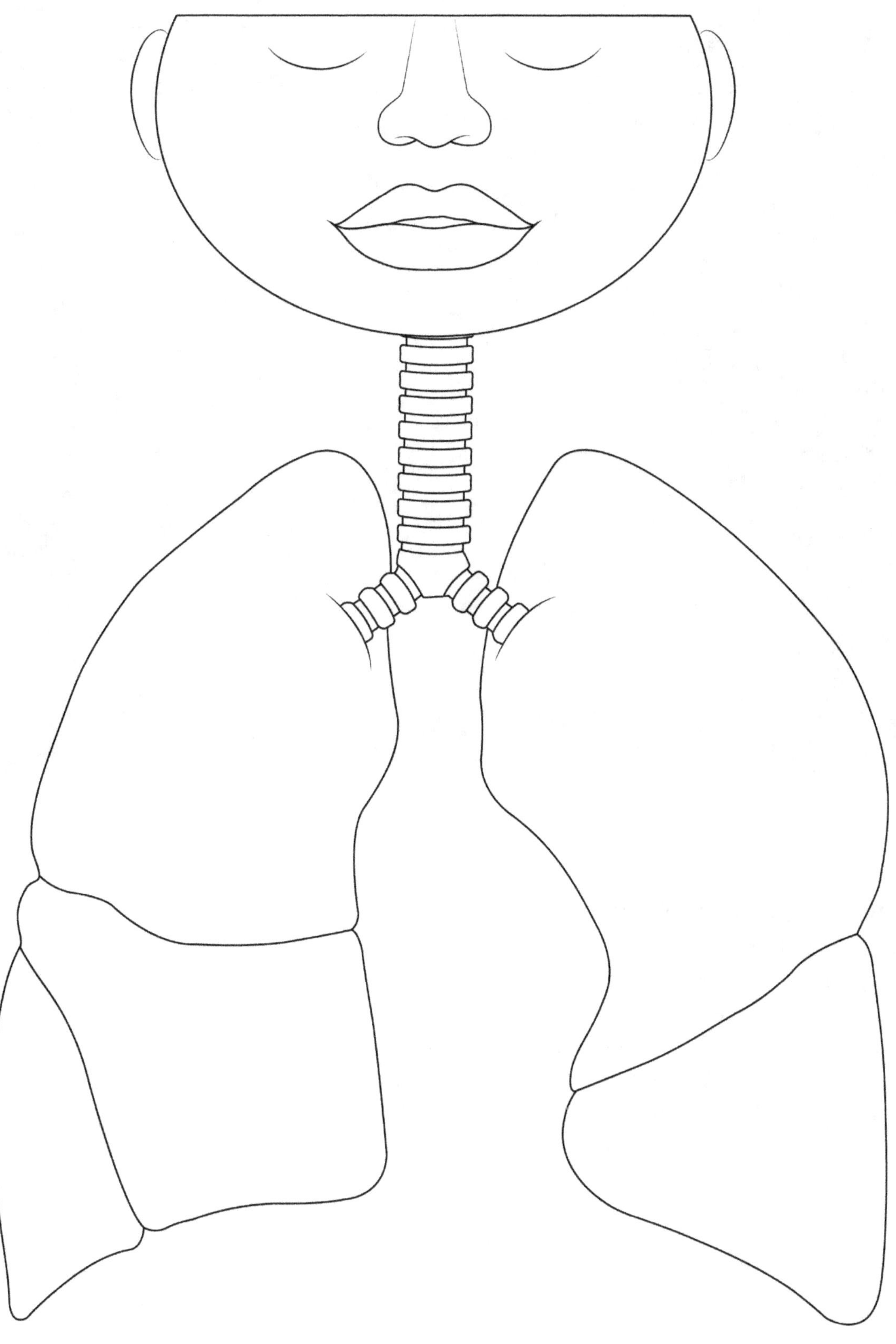

Lungs

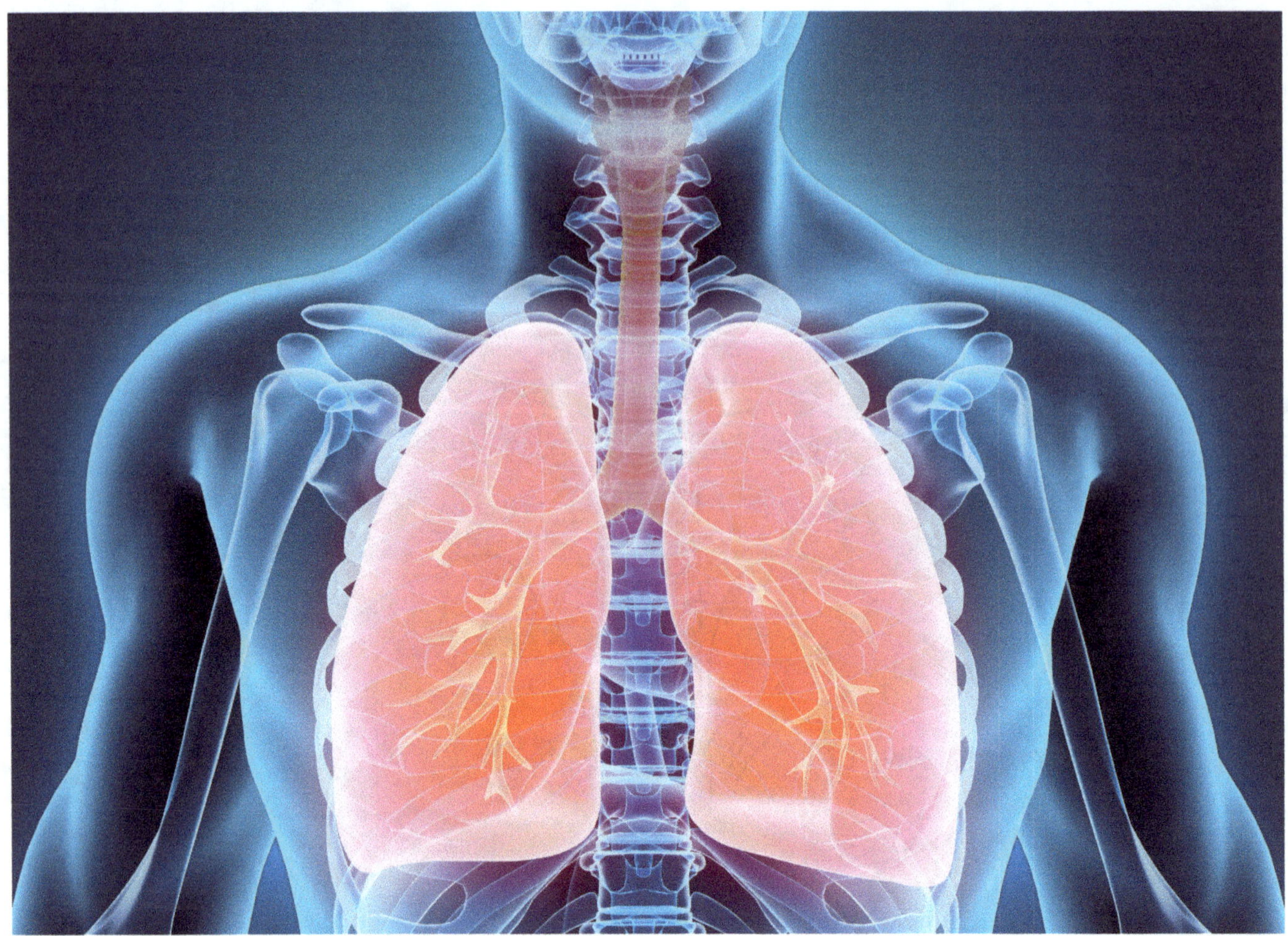

Questions: Respiratory System

1. What word means breathing out?
2. What word means breathing in?
3. Are the esophagus and the trachea (windpipe) the same thing?
4. What is an odorless gas used for energy, growth, and staying alive?
5. What is an odorless, colorless gas the body considers as waste?
6. What is the name of the organs the body uses to breathe air?

Respiratory System
Homework

Directions

Each of these foods strengthen your respiratory system.
Circle the word if you eat or drink one of these foods this week.

Beets

Peppers

Apples

Pumpkins

Turmeric

Tomatoes

Blueberries

Green tea

Red cabbage

Olive oil

Oysters

Yogurt

Brazil nuts

Coffee

Swiss chard

Barley

Lentils

Cocoa

The greatest wealth is health!

Chapter 11
Circulatory System

For the life of the flesh is in the blood…
(Leviticus 17:11a)

Vocabulary

Circulatory system: a system used to deliver nutrients and oxygen to all the cells in the body

Blood: a liquid that carries nutrients and oxygen throughout the body

Heart: the muscle responsible for pumping blood throughout the body

Aorta: the largest artery in the body

Arteries: blood vessels that take blood away from the heart

Capillaries: the smallest blood vessels that connect arteries to veins

Veins: the blood vessels that take blood back to the heart

Vena cava: large veins (inferior and superior) that carry blood back into the heart

Lungs: the organs the body uses to breathe air

Nutrients: fats, proteins, carbohydrates, vitamins, and minerals

Pulse: the thump in the arteries whenever the heart beats

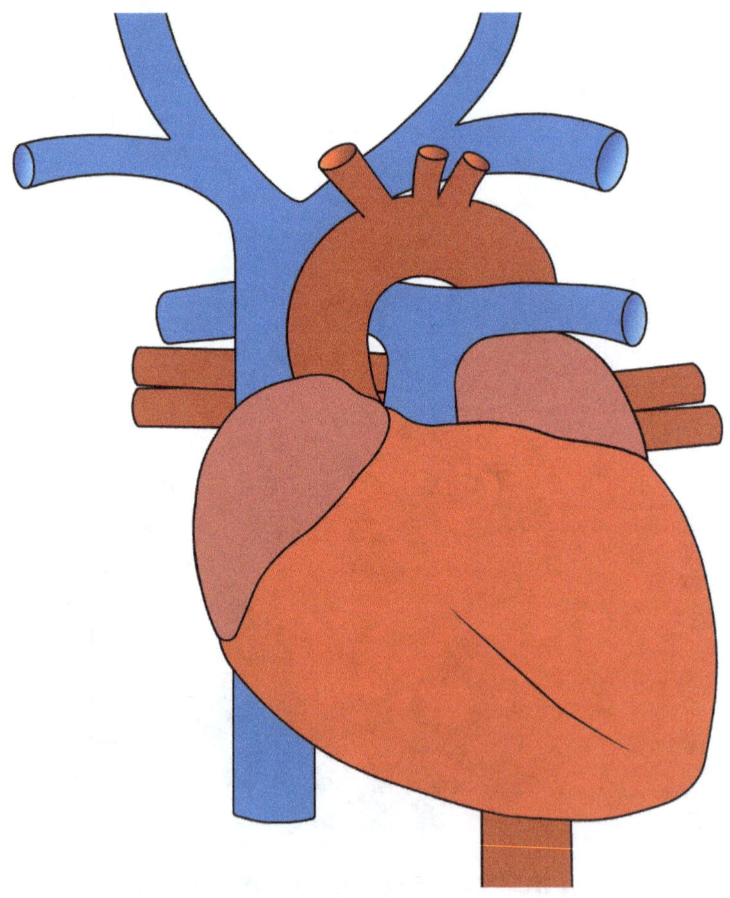

If all your blood vessels were laid end to end, they would stretch about 60,000 miles. This is about 2.5 times around the world.

The Circulatory Journey

Follow the Arrows
Red represents blood with oxygen.
Blue represents blood without oxygen.

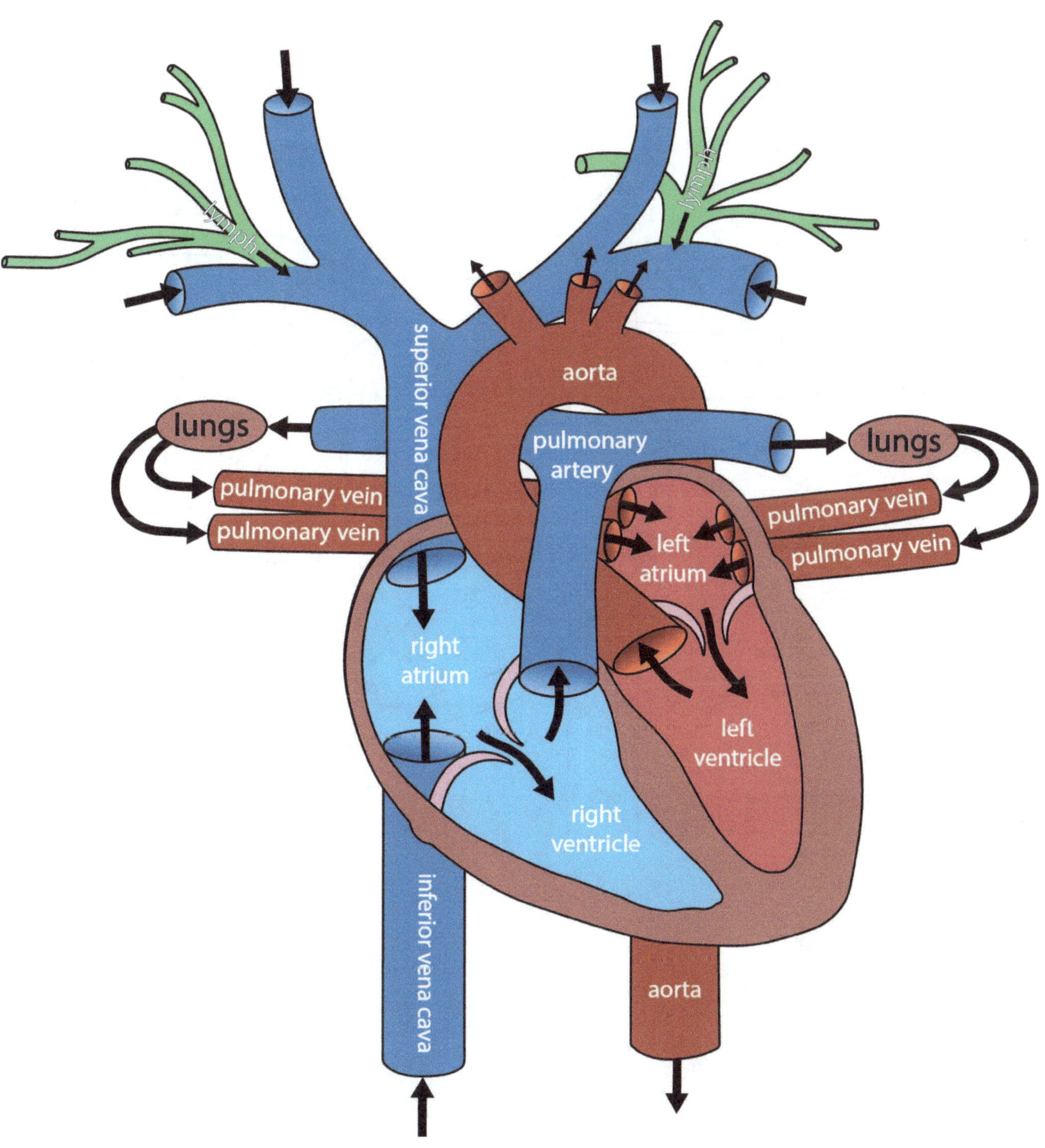

Heart

lymph vessels
aorta
lymph vessels
pulmonary artery
superior vena cava
pulmonary vein
pulmonary vein
left atrium
right atrium
left ventricle
right ventricle
inferior vena cava
aorta

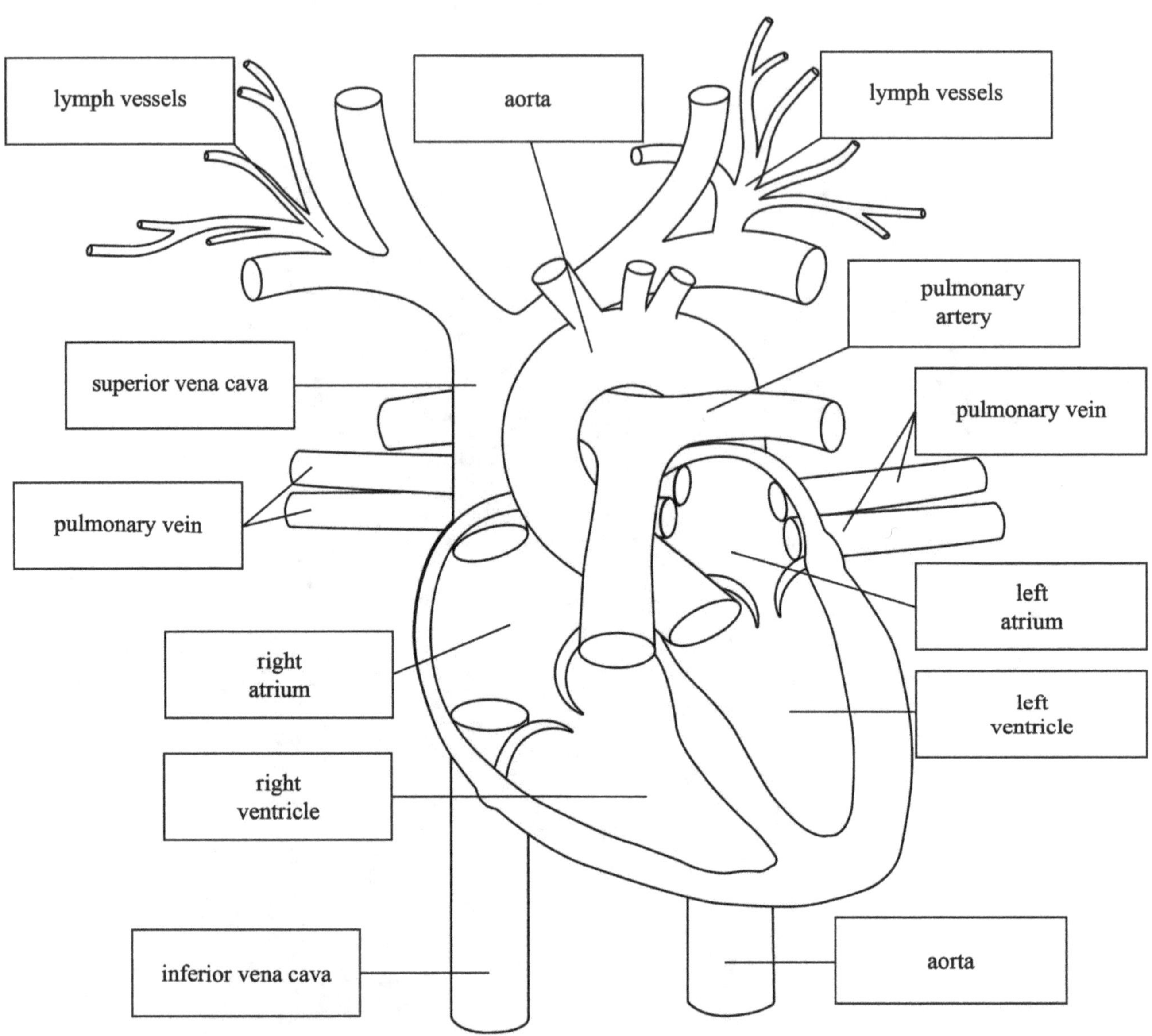

Heart

Label and color the diagram by placing the correct word, or words, in the boxes below.
See page 58-59 for color and placement.

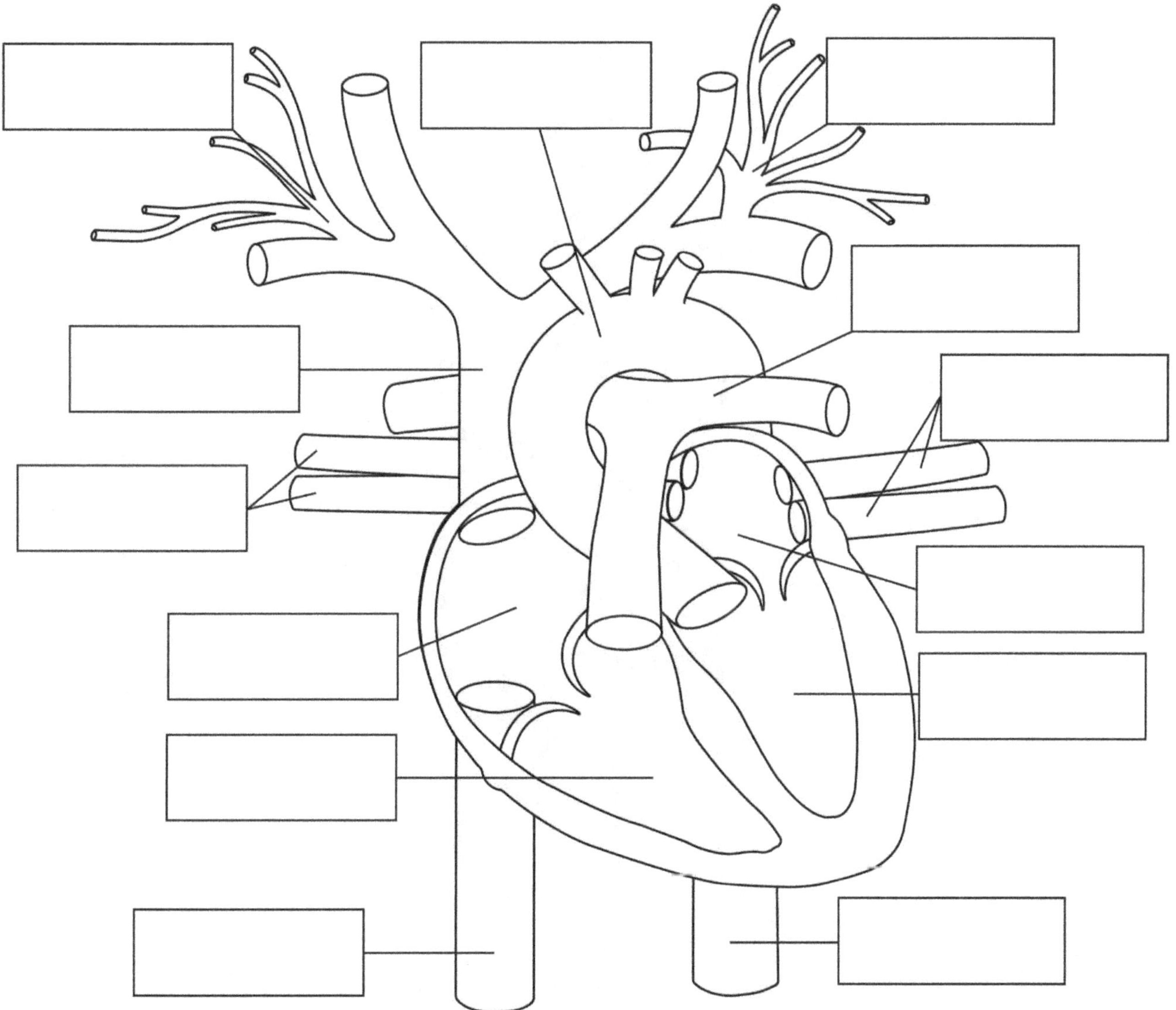

1. aorta
2. lymph vessels
3. inferior vena cava
4. superior vena cava
5. pulmonary artery
6. pulmonary vein
7. left atrium
8. right atrium
9. left ventricle
10. right ventricle

The Circulatory Journey

Follow the Arrows

Red represents blood with oxygen.
Blue represents blood without oxygen.

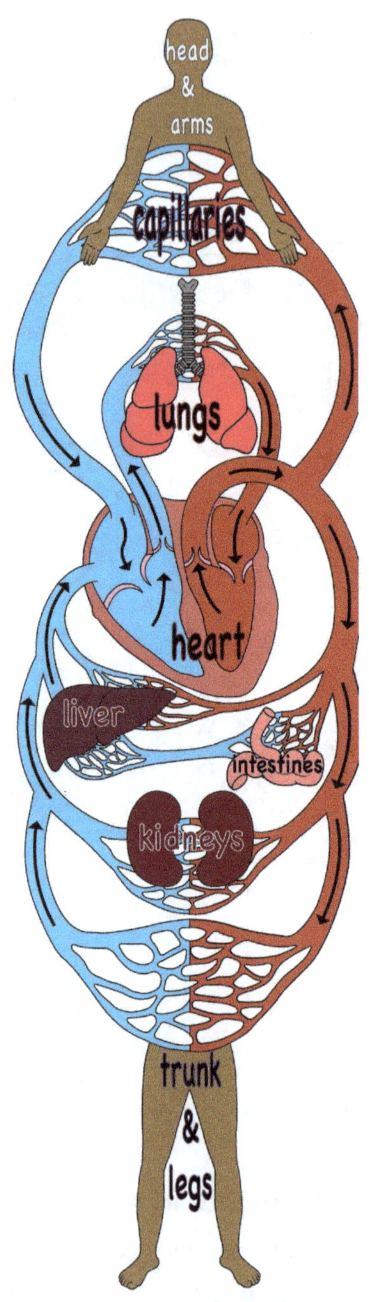

A Never-Ending Circle

Blood is pumped from the heart to travel throughout the body.
Blood collects nutrients from the intestines.
Blood travels from the intestines to the liver for processing.
Blood proceeds through the kidneys to be filtered and cleaned.
Blood advances to capillaries to deliver oxygen and nutrients and pick up waste.
Blood returns to the heart and then is sent to the lungs to gather oxygen.
Blood journeys back to the heart.
The cycle begins again!

Pulse
Activity

Purpose
To learn the proper way to check your pulse.

Materials
Just you!
Optional: Finger pulse oximeter

Procedure
Place the pads of your first and middle fingers on the side of you neck or on the inside of your wrist just below the base of your thumb. Press lightly to feel your pulse for fifteen seconds.
Put your results on the chart below.

Result
You should feel a slight thumping. That is blood moving through your arteries!

Why?
Your pulse shows your heart is beating and your circulatory system is working.

	Heart Rate per 15 Seconds		Beats per Minute
Resting:		**X4=**	
After Workout:		**X4=**	
Recovery:		**X4=**	

Circulatory System

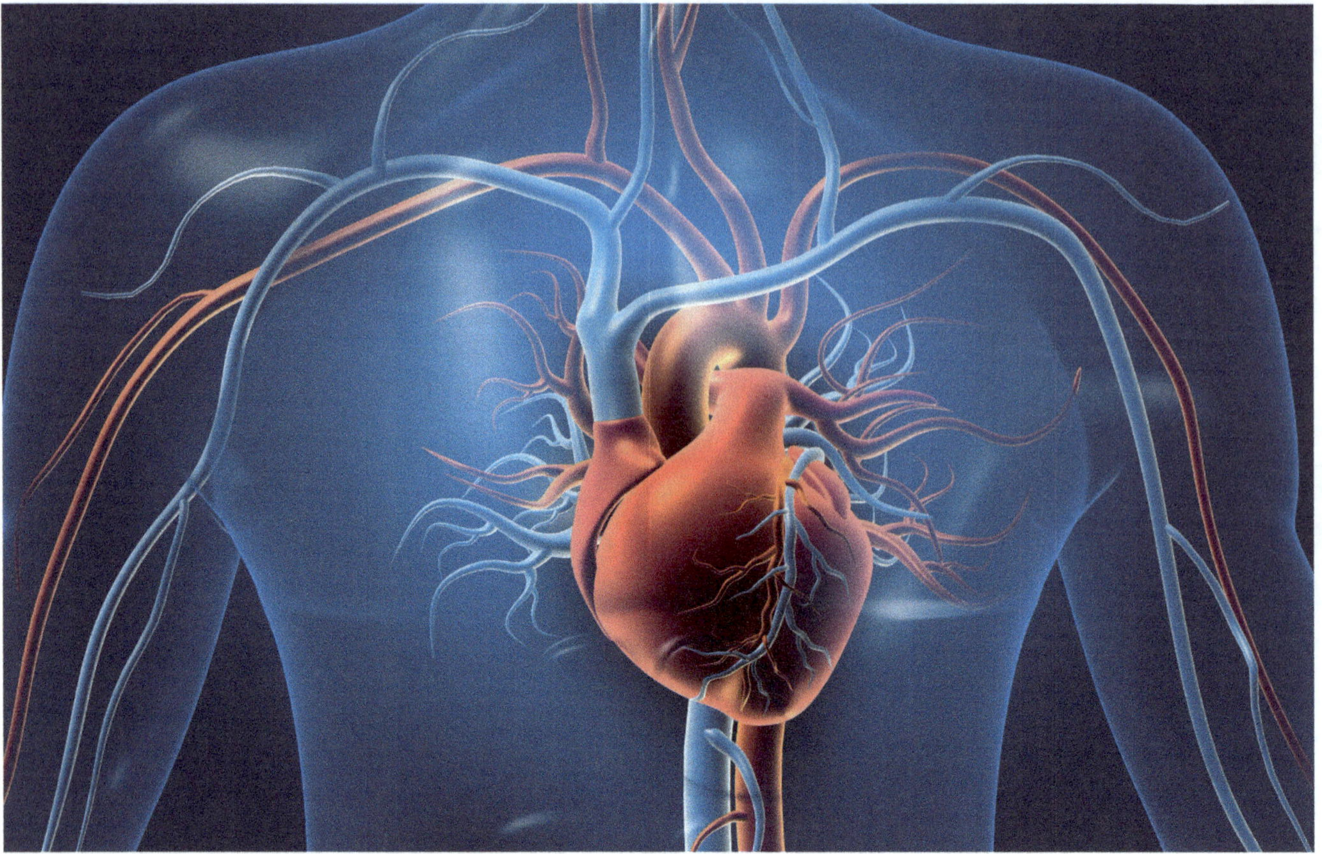

Questions: Circulatory System

1. What body system delivers nutrients and oxygen to the body's cells?
2. What organ pumps blood throughout the body?
3. Do arteries bring blood away from the heart or back to the heart?
4. What is the largest artery in the body?
5. What color is the blood in most arteries?
6. Do veins bring blood away from the heart or back to the heart?
7. What are the two largest veins in the body?
8. What color is the blood in most veins?
9. If all your blood vessels were spread out, how many miles would they stretch?
10. What are the smallest blood vessels called?
11. What organ puts oxygen into blood?

Circulatory System
Homework

Directions

Each of these foods strengthen your circulatory system.
Circle the word if you eat or drink one of these foods this week.

Garlic

Cayenne pepper

Turmeric

Ginger

Nuts

Beets

Onions

Oranges

Lemons

Grapefruit

Blueberries

Blackberries

Cranberries

Fish

Walnuts

Love yourself enough to live a healthy lifestyle!

Chapter 12
Lymphatic System

Or do you not know that your body is a temple of the Holy Spirit within you,
whom you have from God, and that you are not your own?
For you have been bought for a price:
therefore, glorify God in your body.
(1 Corinthians 6:19-20)

Vocabulary

Bone marrow: the place where red blood cells, white blood cells, and platelets are created

Lymphatic System: a system used to fight infections with white blood cells; manages fluid levels

Lymph: fluid in the lymphatic system

Lymphatic vessels: carries lymph and white blood cells

Lymph nodes: bean-shaped glands along lymphatic vessels that filter lymph and fight infections

Spleen: filters blood and stores white blood cells, red blood cells, and platelets

Thymus gland: produces hormones to help with the body's defense against disease

Tonsils: traps bacteria and viruses that enter the mouth and nose

Adenoids: trap bacteria and viruses that enter the mouth and nose

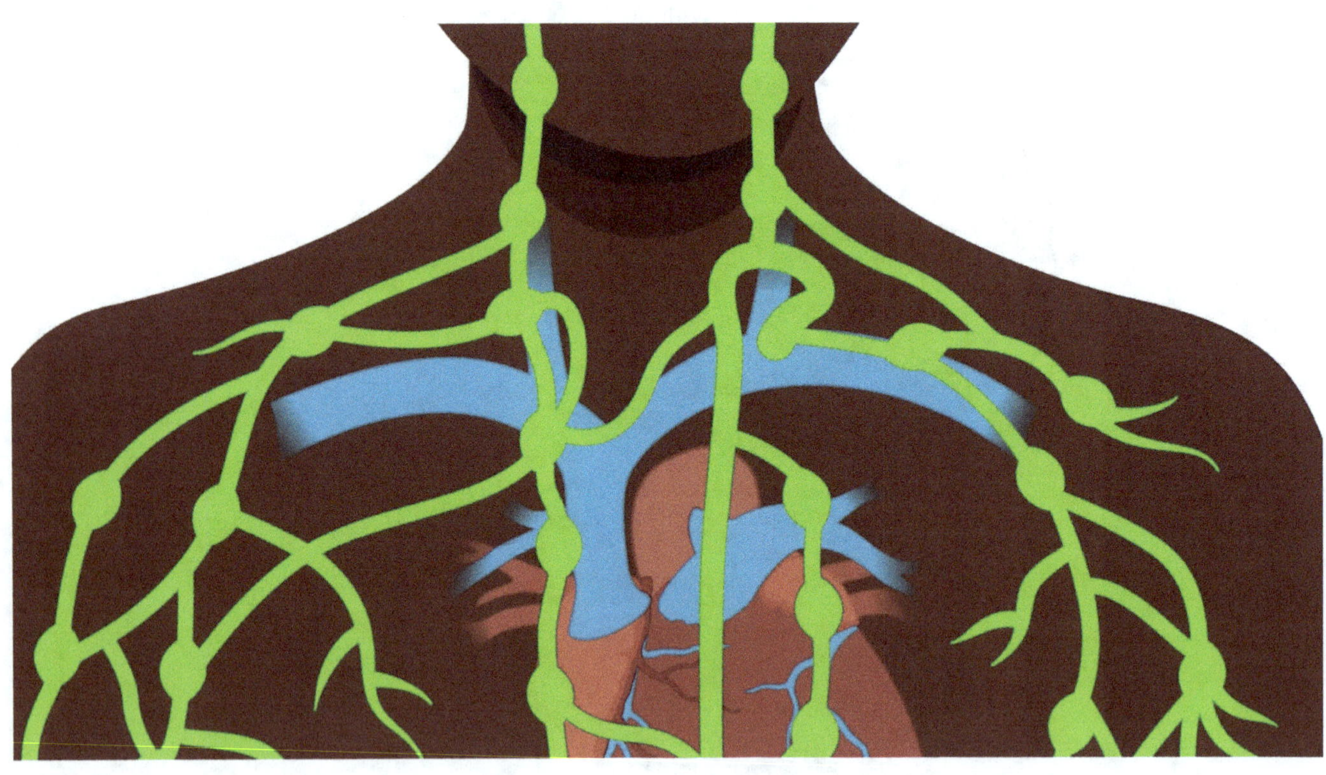

Lymphatic System Exercises

Pelvic Tilt

Lie on your back with your hands behind your head. Bend your knees and place your feet flat on the floor. Tuck your tailbone under, press the small of your back into the floor, hold that position for 3 seconds, and slowly release. Repeat 10 times.

Neck Rotation

As you inhale, slowly turn your head to the right. Count to five. Exhale while bringing your head back to center. Repeat the same thing on the left side. Repeat five times on each side.

Shoulder Shrug

Stand or sit in a comfortable position. As you inhale, draw both shoulders up toward your ears and then exhale, releasing your shoulders to a neutral position. Repeat five times.

Leg Slides

Lie on your back with your arms alongside your body and your legs straight in front of you. Inhale, sliding your right leg out to the side. Exhale while sliding your right leg back to center. Next, slide your left leg out to the side and bring it back to center. Repeat five times with each leg.

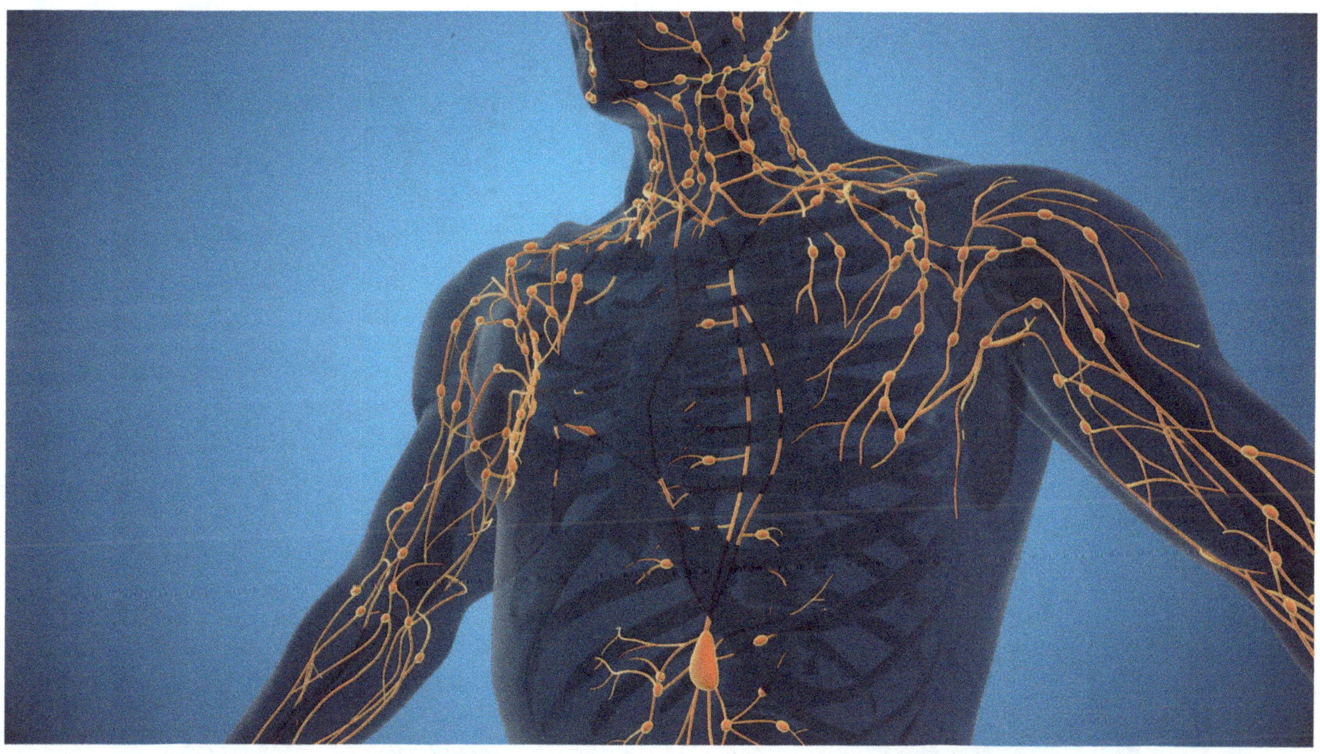

Exercise releases blockages and promote healthy movement of nutrients and waste.

Lymphatic Activity

Unscramble the words or fill in the blank.

1. LISNOT
2. RROWMA
3. DOOLB
4. NEELPS
5. LYMHP
6. Lymphatic vessels carry lymph and white blood __________.
7. Lymph is ______________ in the lymphatic system.
8. The spleen filters ________________ and stores white blood cells.
9. Red and white blood cells are created in the ___________ ______________.
10. ______________ is good for the lymphatic system.

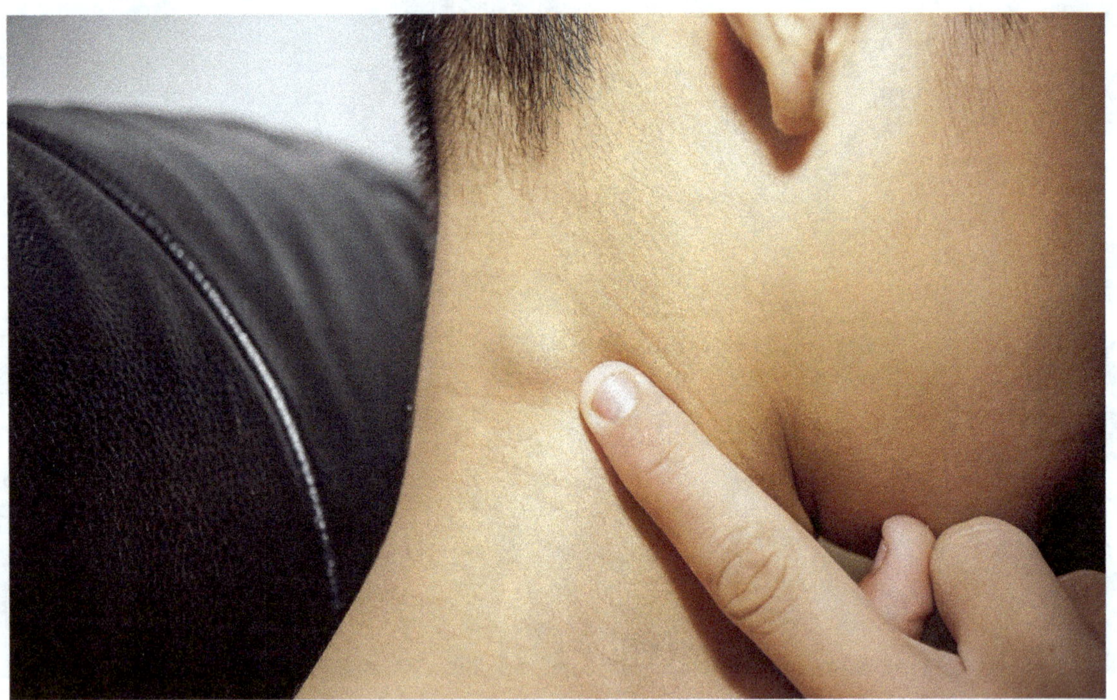

Lymph nodes swell when you are sick.

Lymphatic System

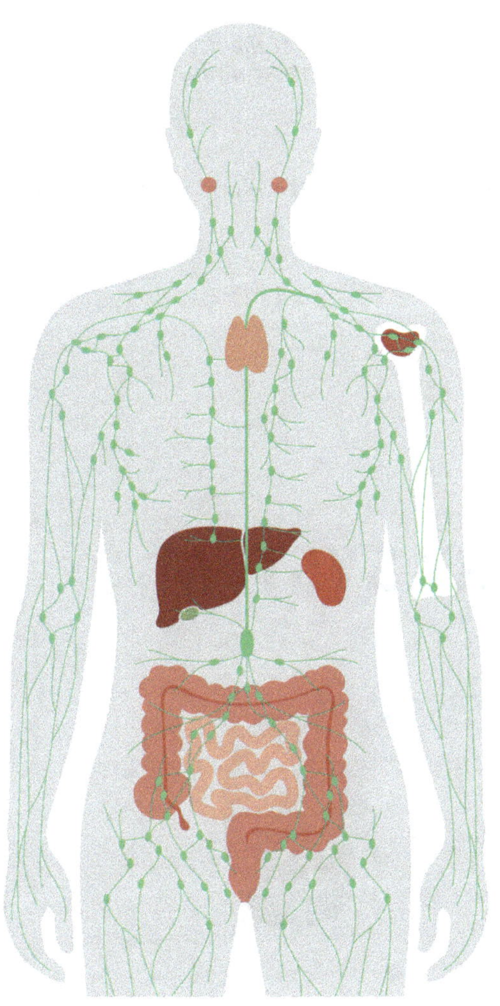

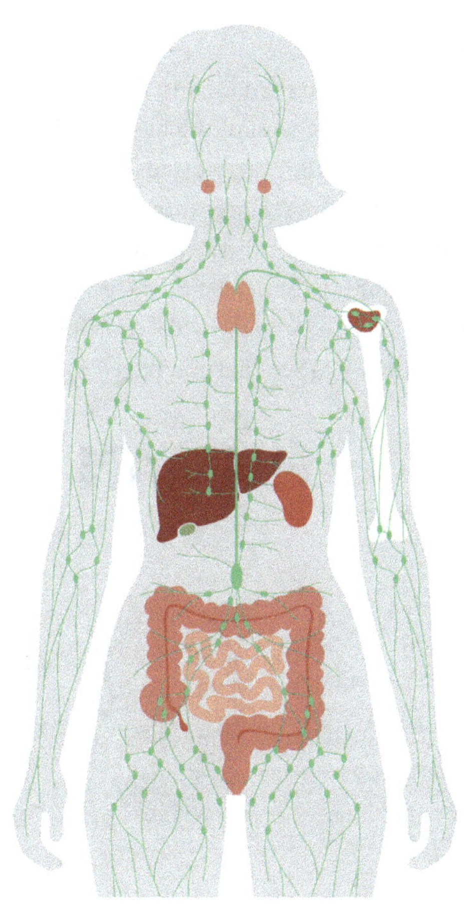

Questions: Lymphatic System

1. What two organs trap bacteria and viruses entering the nose and mouth?
2. What are the jobs of the lymphatic system?
3. What organ filters blood and stores white blood cells?
4. What kind of nodes drain lymph fluid?
5. What gland releases hormones to help with the body's defense against disease?

Chapter 13
Urinary System

Then take all the fat on the internal organs, the long lobe of the liver, and both kidneys with the fat on them, and burn them on the altar.
(Exodus 29:13)

Vocabulary

Urinary System: a system used to filter waste and remove excess fluid from the blood in the form of urine

Kidneys: filters waste out of the blood and gets rid of it as urine

Ureters: the tubes that carry urine from the kidneys to the bladder

Bladder: an organ that stores urine

Urethra: the tube that rids the body of urine

Acidic: a pH of 0-6 (See chart on page 71)

Alkaline: a pH of 8 or higher (See chart on page 71)

Neutral: a pH of 7 (See chart on page 71)

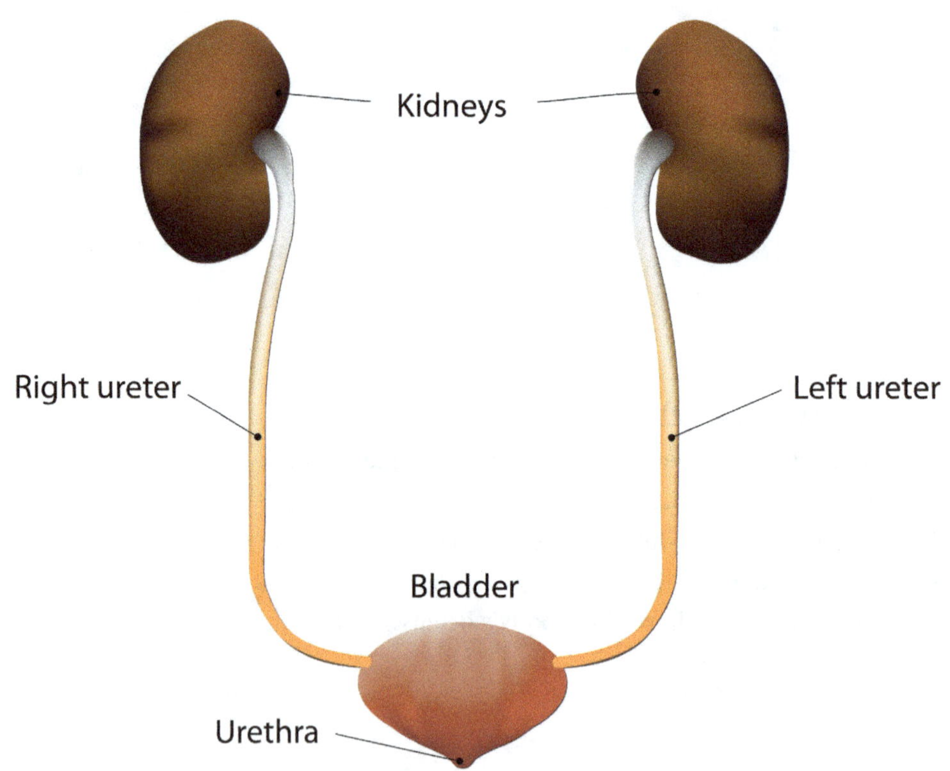

Alkaline Versus Acidic Foods

Scientists measure pH levels on a scale of 0-1

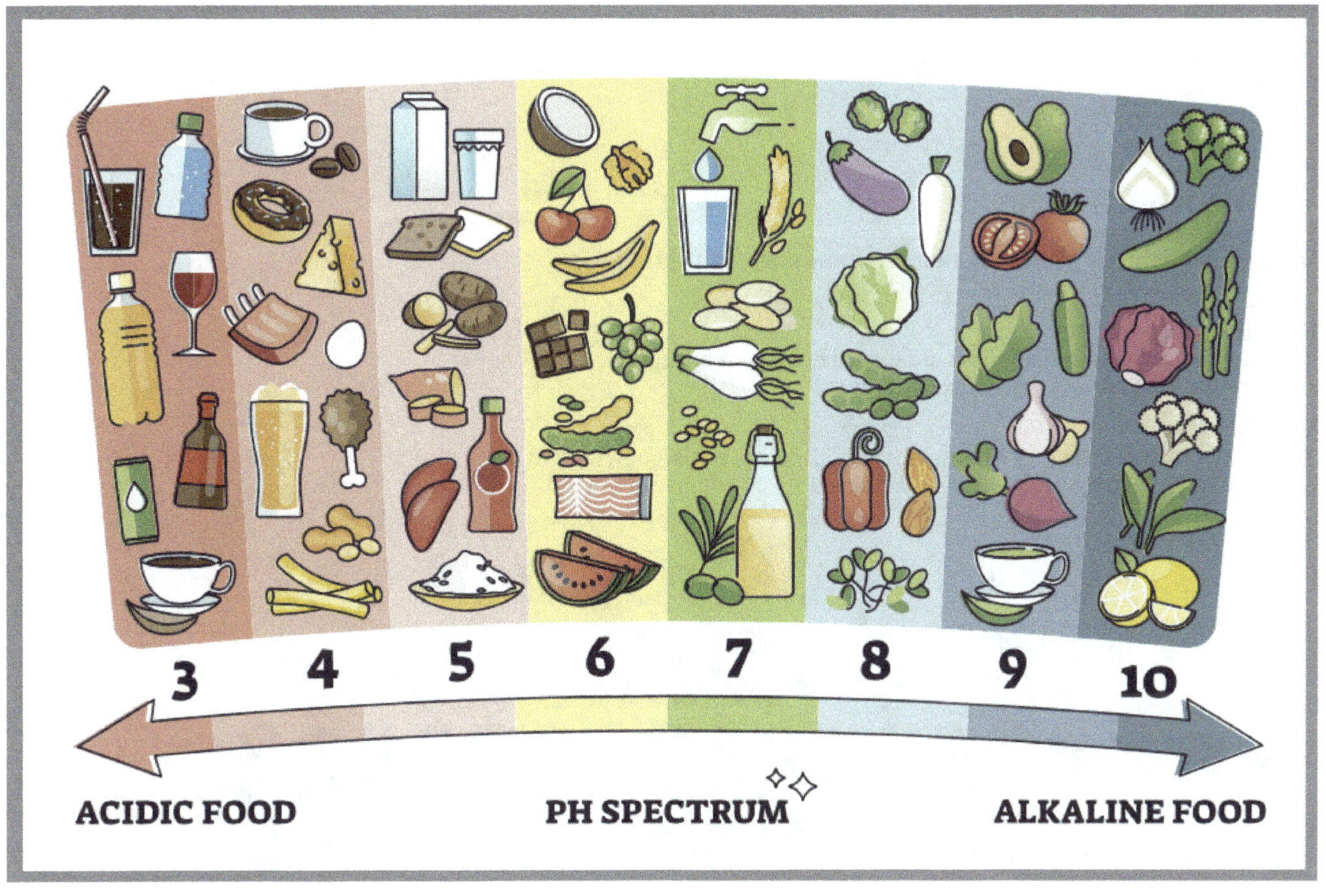

When your body becomes too acidic or too alkaline, it can lead to various symptoms including sleepiness, fatigue, headache, loss of appetite, confusion, and vomiting. A healthy pH is slightly alkaline - average 7.40.

pH Scale
Activity

Purpose
To encourage a healthy lifestyle

Materials
pH scale
pH strips
saliva (spit)

Procedure
Spit on a pH strip.
Wait for it to change color.
Determine if you are acidic, alkaline, or somewhere in the middle.
Circle your color on the pH scale below.

Result
A pH below seven is considered acidic, a pH of seven is neutral, and anything above 7 is considered alkaline.

Why?
The goal is to be neither too acidic nor too alkaline, but somewhere in the middle. A healthy pH number is achieved by eating healthy food.

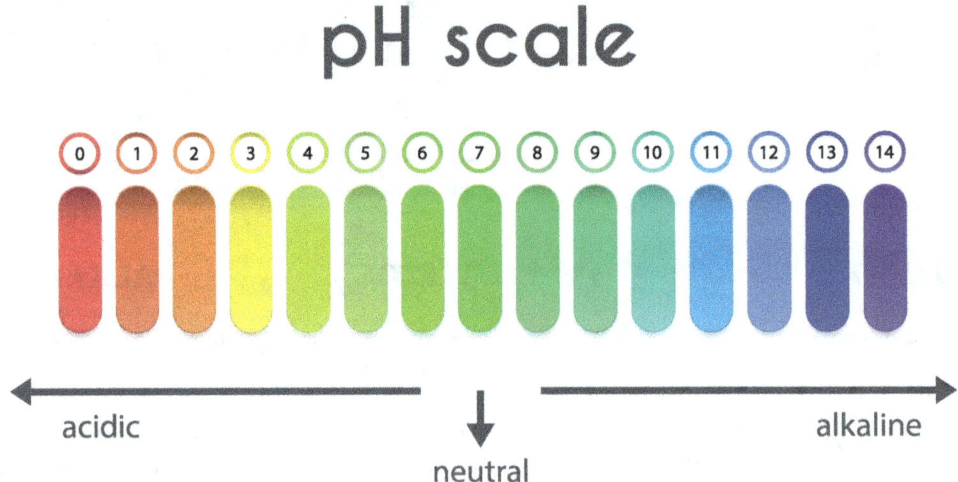

Using saliva and pH strips, what number did you score on the pH scale?

Cranberries

Cranberries help prevent urinary tract infections.

Questions: Urinary System

1. What organ stores urine?
2. What organ filters body fluids and removes liquid waste?
3. What tubes carry urine from the kidneys to the bladder?
4. What tube rids the body of urine?

Urinary System
Homework

Water helps the kidneys remove waste from your blood in the form of urine.
Drink 6 to 8 glasses of water each day (8 ounces)

Put an x in the boxes for every glass of water you drink.
(Each glass should hold 6-8 ounces of water.)

MONDAY								
TUESDAY								
WEDNESDAY								
THURSDAY								
FRIDAY								

Directions
Each of these foods strengthen your kidneys.
Circle the foods you eat or drink this week.

Red peppers
Berries
Fish
Bananas
Oranges
Tomatoes
Avocado
Broccoli
Nuts
Squash
Lemon
Cranberry

Eat Healthy!

Chapter 14
Jeopardy Review

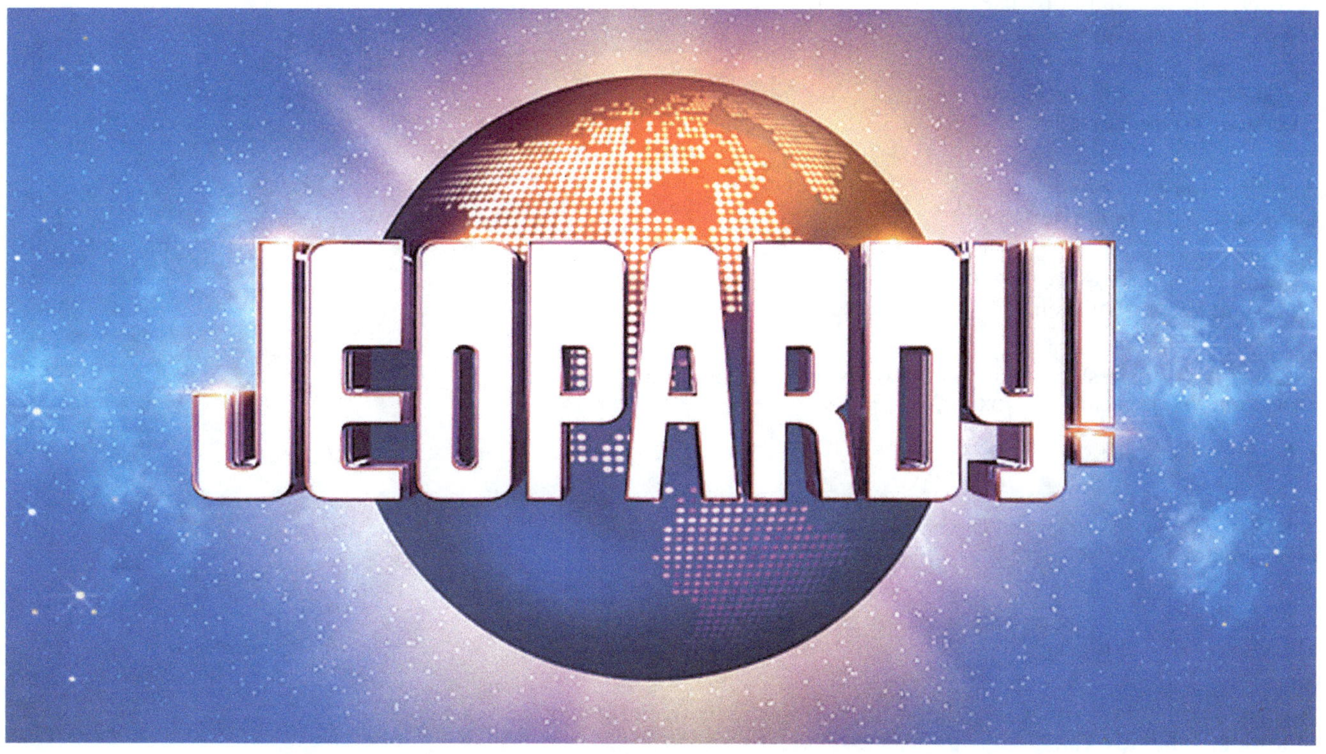

Goal: To create **Questions** from **Answers** instead of **Answers** from **Questions**

On a white board, write the categories and point values provided below.

Divide the students into teams.
Use the **Answers** and **Questions** found on pages 75-76.
The teams will take turns until all the **Answers** and **Questions** are used.
Rotate students so everyone gets a turn.
Keep team scores on a white board.

Procedure: A student from each team will choose a category and a point amount.
The teacher will read the **Answer** connected to this choice.
This student will try to think of a **Question** that goes with the **Answer.**

If the student comes up with the correct **Question,**
he or she will receive full point value.
If the student cannot come up with the **Question,**
then ALL teams will concur with each other in separate parts of the room.

ALL the teams with the correct **Question** connected to the **Answer,** will get half the point value.
The **Question** can be written or whispered into the teacher's ear.

The game continues until all the categories and point values
have been chosen and the correct **Questions** have been provided.
The team with the most points wins.

Respiratory System	Circulatory System	Lymphatic System	Urinary System
10	10	10	10
20	20	20	20
30	30	30	30
40	40	40	40
50	50	50	50

Respiratory System

10
Answer: The organ a body uses to breathe air
Question: What are lungs?

20
Answer: Means to breathe out
Question: What is exhale?

30
Answer: Means to breathe in
Question: What in inhale?

40
Answer: An odorless gas used for energy, growth, and staying alive
Question: What is oxygen?

50
Answer: An odorless colorless gas the body needs to eliminate
Question: What is carbon dioxide?

Circulatory System

10
Answer: The largest artery in the body
Question: What is the aorta?

20
Answer: The liquid that carries nutrients and oxygen throughout the body
Question: What is blood?

30
Answer: Blood vessels that take blood back to the heart from the body
Question: What are veins?

40
Answer: Fats, proteins, carbohydrates, vitamins, and minerals
Question: What are nutrients?

50
Answer: The smallest blood vessels that connect arteries to veins
Question: What are capillaries?

Lymphatic System

10
Answer: The system that manages the body's fluids
Question: What is the lymphatic system?

20
Answer: Traps bacteria and viruses that enter the mouth and nose
Question: What are adenoids? or What are tonsils?

30
Answer: Releases hormones to help with the body's defense against disease
Question: What is the thymus gland?

40
Answer: Filters blood and stores white blood cells
Question: What is the spleen?

50
Answer: Releases hormones to help with a body's defense against disease
Question: What is the thymus gland?

Urinary System

10 Points
Answer: The organ in the body that stores urine
Question: What is the bladder?

20 Points
Answer: Measured with spit using pH strips
Question: What is alkaline and acidic?

30 Points
Answer: When the pH is 8 or higher: alkaline or acidic
Question: What is alkaline?

40 Points
Answer: When the pH is 0-6: alkaline or acidic
Question: What is acidic?

50 Points
Answer: Filters waste out of the blood and gets rid of it as urine
Question: What are kidneys?

Chapter 15
Digestive System
Overview

Don't you see that whatever enters the mouth goes into the stomach and then out of the body? (Matthew 15:17)

Vocabulary

Digestive System: a system used to break down food for energy, growth, and tissue repair

Mouth: the beginning of the digestive system where food enters the body

Salivary glands: produce saliva to break down food

Esophagus: a tube used to carry food and liquid from the throat to the stomach

Stomach: mixes, liquifies, and moves food using acids and enzymes

Small intestine: sends nutrients to cells via the bloodstream

Large intestine: dehydrates unused food and removes it from the body as waste material

Liver: cleanses blood and helps digestion by secreting bile

Gallbladder: stores bile produced by the liver

Bile: a liquid that breaks down fats

Pancreas: creates hormones to regulate blood sugar levels in blood

Dehydrate: to remove water

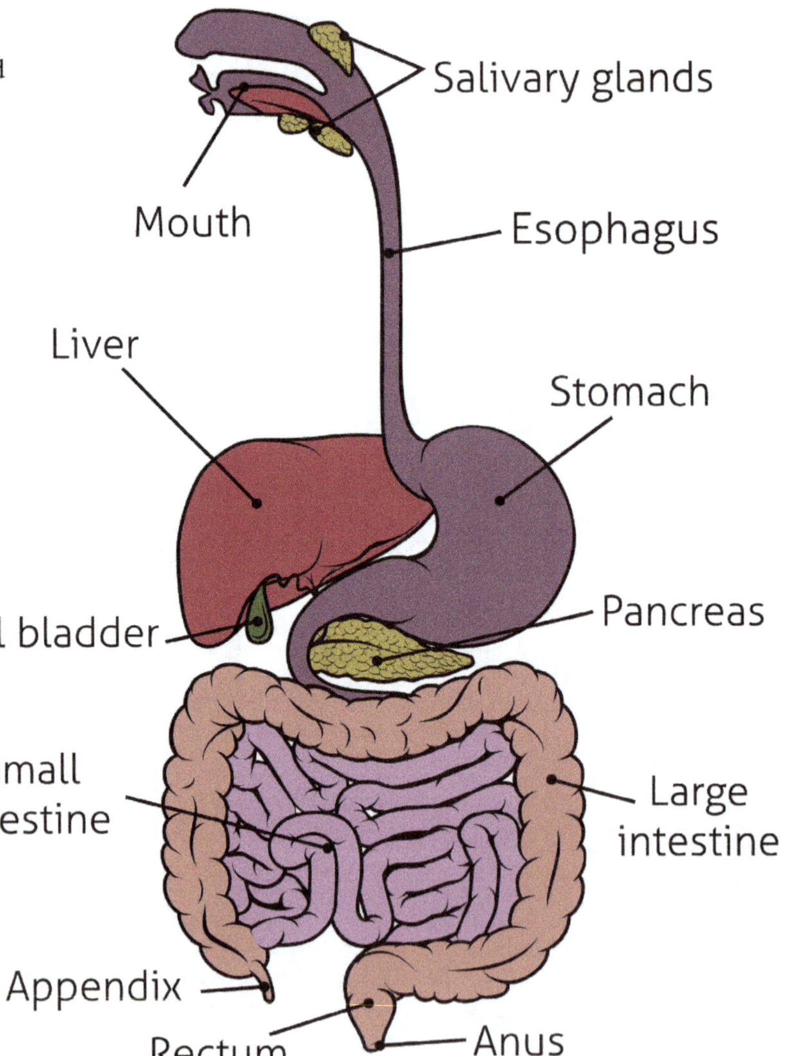

DIGESTIVE SYSTEM
WORD SEARCH
Answers on page 138

B	F	P	Q	S	E	N	I	T	S	E	T	N	I
A	P	P	E	N	D	I	X	J	R	T	A	B	P
C	D	E	S	Z	V	I	H	O	A	S	D	A	F
G	S	E	W	A	G	H	J	K	L	Z	R	X	C
R	T	H	H	X	L	I	V	E	R	C	E	B	N
S	O	S	Q	Y	M	I	Q	W	R	E	D	R	E
Y	M	U	P	I	D	Y	V	E	U	I	D	E	V
T	A	G	Y	Z	P	R	A	A	Z	X	A	C	I
U	C	A	S	V	N	S	A	B	R	N	L	T	T
H	H	H	A	S	K	D	E	T	F	Y	B	U	S
T	G	P	H	J	K	S	L	A	E	S	L	M	E
U	F	O	G	L	M	U	H	J	K	U	L	D	G
O	W	S	N	E	R	N	T	Y	U	I	A	C	I
M	O	E	P	A	N	C	R	E	A	S	G	W	D

GALLBLADDER	DEHYDRATE	ESOPHAGUS
APPENDIX	MOUTH	PANCREAS
INTESTINES	STOMACH	RECTUM
SALIVARY	LIVER	DIGESTIVE

Digestive System

Questions: Digestive System

1. Where is the starting place of digestion?
2. The salivary glands produce what kind of fluid?
3. What tube connects the mouth to the stomach?
4. What organ mixes and breaks down food into liquid using acids and enzymes?
5. What organ sends nutrients to the cells?
6. What organ dehydrates and sends unused food out as waste material?
7. What does "dehydrate" mean?
8. What organ produces bile and cleanses blood?
9. What organ stores bile?
10. What does bile do?

Digestive System
Homework

Directions

Each of these foods strengthen your digestive system.
Circle the word if you eat or drink one of these foods this week.

Yogurt
Apples
Fennel
Kefir
Chia seeds
Kombucha
Papaya
Oats
Quinoa
Beets
Ginger
Almonds
Liver
Salmon
Tomatoes
Broccoli

Eat well!
Live well!
Be well!

Chapter 16
Mouth and Esophagus
Digestive System

The sluggard buries his hand in the dish,
But will not even bring it back to his mouth.
(Proverbs 19:24)

Vocabulary

Mouth: the beginning of the digestive system where food enters the body

Teeth: small, white structures in the mouth used for chewing

Enamel: a hard, white substance that covers the surface of teeth

Tongue: a muscle in the mouth used for tasting, licking, swallowing, and speaking

Salivary glands: produces saliva to break down food

Saliva: a clear liquid in the mouth that helps begin digestion, also known as spit

Esophagus: a tube used to carry food and liquid from the throat to the stomach

Teeth
Activity

Purpose
To learn the value of taking care of your teeth

Materials
5 clear glasses
5 eggs
Water
Apple juice
Cola
Lemon juice
Vinegar
Toothpaste
Toothbrush

Procedure
Place an egg in each glass.
Fill each glass with different liquids.
Let the eggs sit for 48 hours.
Put each egg on a plate.
Observe the results.
Brush the discolored eggs with a toothbrush and toothpaste.

Result
The egg in the water will be unchanged.
The egg in the apple juice will be discolored.
The egg in cola will be discolored.
The egg in the coffee will be unchanged.
The egg in the vinegar will no longer have a shell and will be rubbery.

Why?
The chemical reactions determine what will happen to the eggs.
Vinegar removes the shell from the egg because of its acidic content.

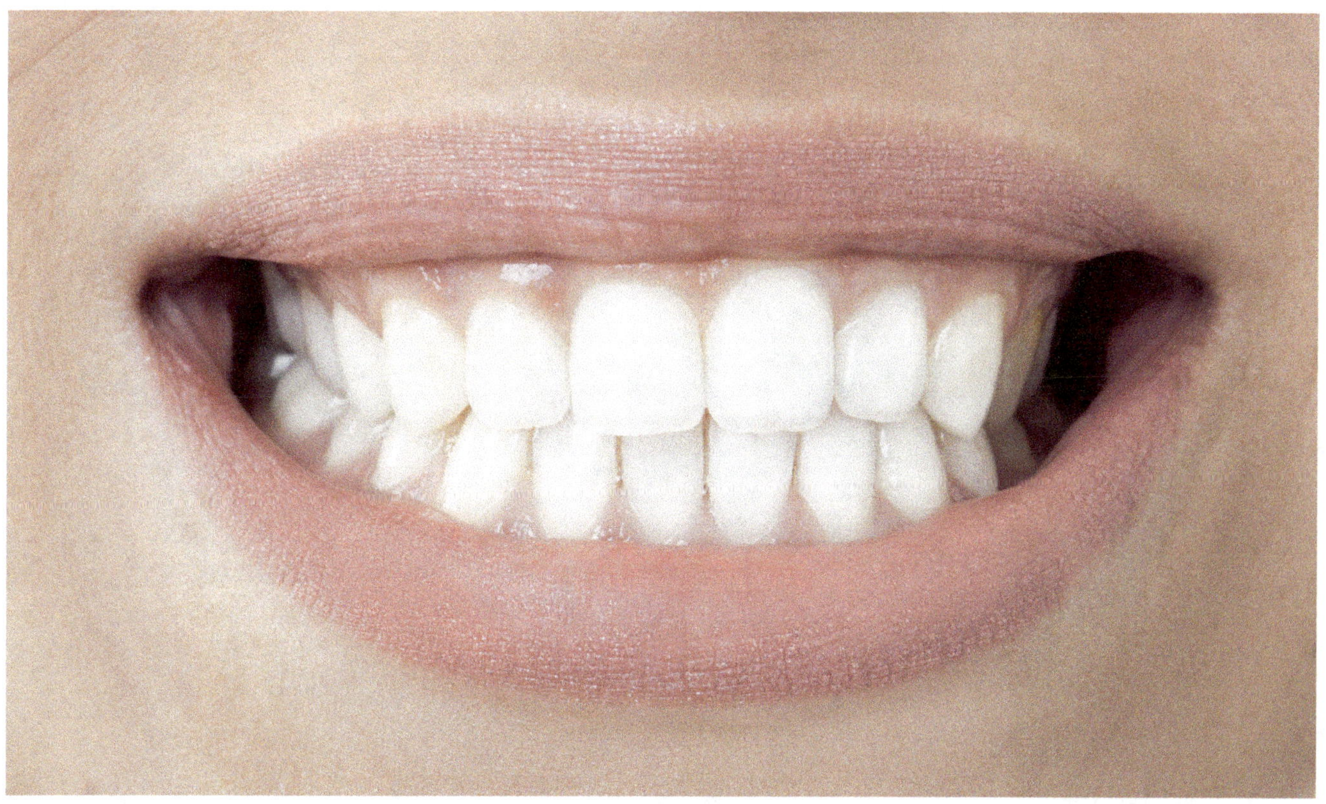

Mouth and Esophagus

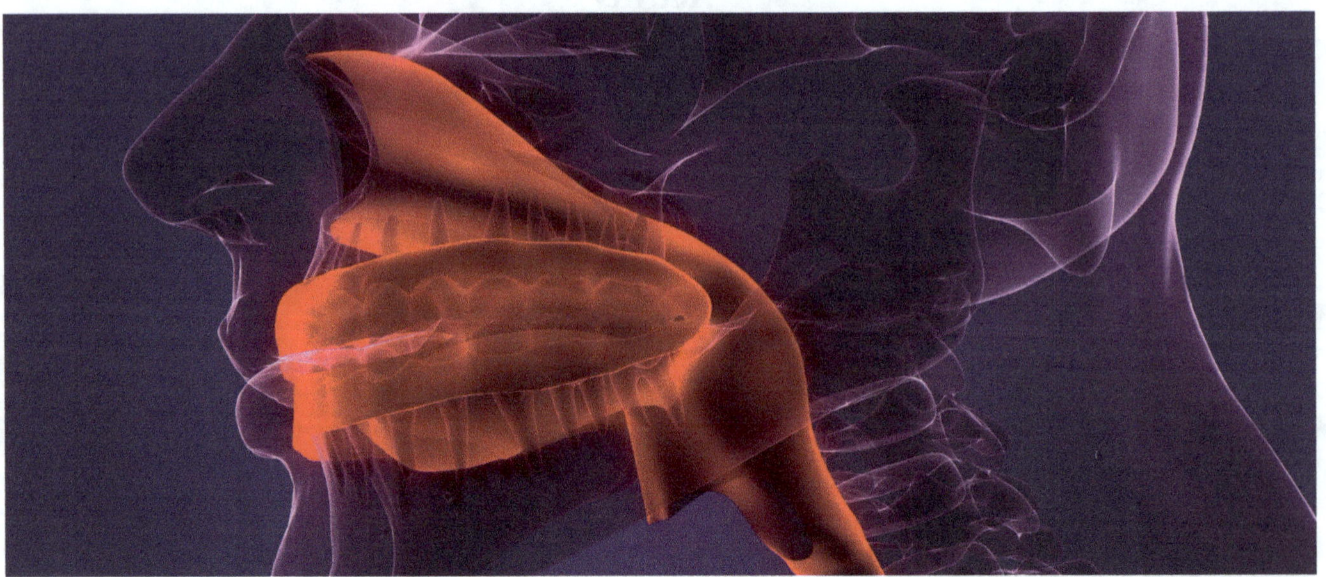

Questions: Mouth and Esophagus

1. What part of the body starts the process of breaking food into smaller parts?
2. Name three parts in the mouth used in digestion?
3. What clear liquid in the mouth helps with digestion?
4. What muscle in the mouth pushes food into the throat?
5. What hollow tube goes down the throat to the stomach?
6. What substance covers the surface of teeth?
7. What are some of the ways to take care of teeth?

Mouth and Esophagus
Homework

Directions

Each of these foods strengthen your teeth.
Circle the word if you eat or drink one of these foods this week.

Carrots
Celery
Broccoli
Onions
Spinach
Lettuce
Kale
Apples
Fish
Kale
Strawberries
Pineapple
Ginger
Almonds
Liver
Salmon
Tomatoes
Broccoli

**To be healthy isn't a goal.
It is a way of living.**

Chapter 17
Stomach
Digestive and Muscular Systems

(Do you not understand that everything that goes into the mouth passes into the stomach, and is eliminated?
(Matthew 15:17)

Vocabulary

Stomach: mixes, liquifies, and breaks down food using acids and enzymes

Acids: watery, colorless fluid used by the stomach to break down food

Enzymes: a type of protein used by the stomach to break down food

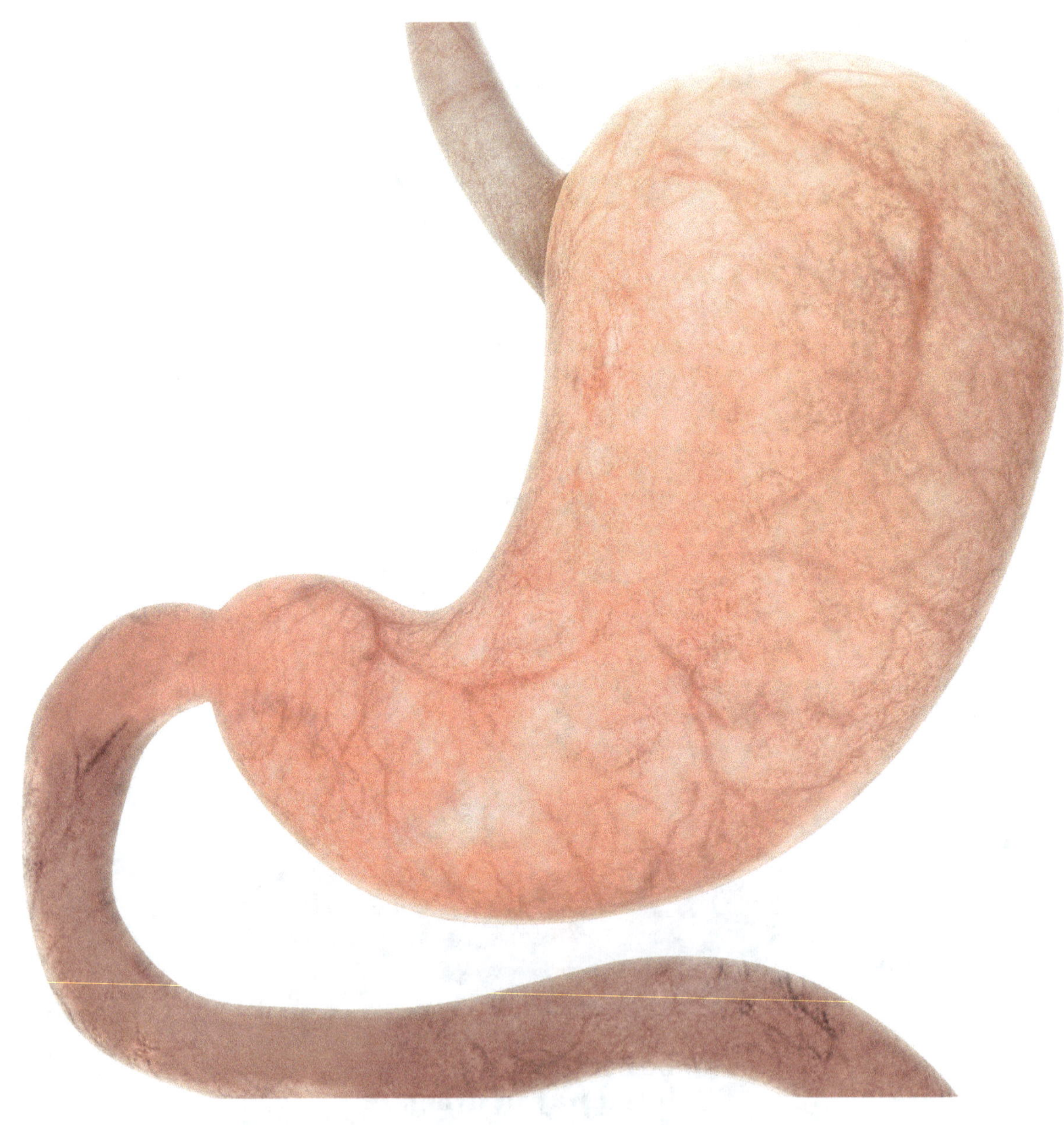

Stomach
Activity

Purpose
To demonstrate how the stomach breaks down solid food into liquid

Materials
White bread
1/2 cup water
1/2 cup apple cider vinegar
Small plastic bags

Procedure
Bread: represents food
Water: represents saliva
Apple cider vinegar: represents stomach acids and enzymes

Put all ingredients in a small plastic bag and seal it.
Squish the ingredients together until it all becomes liquid.

Why?
The stomach uses acids and enzymes to turn food into liquid.
This experiment uses apple cider vinegar (acid) mixed with water (saliva) to demonstrate what happens to our food inside of our stomachs.

It takes between 40 minutes and 2 hours for the stomach to empty its contents into the small intestines for processing.

Stomach

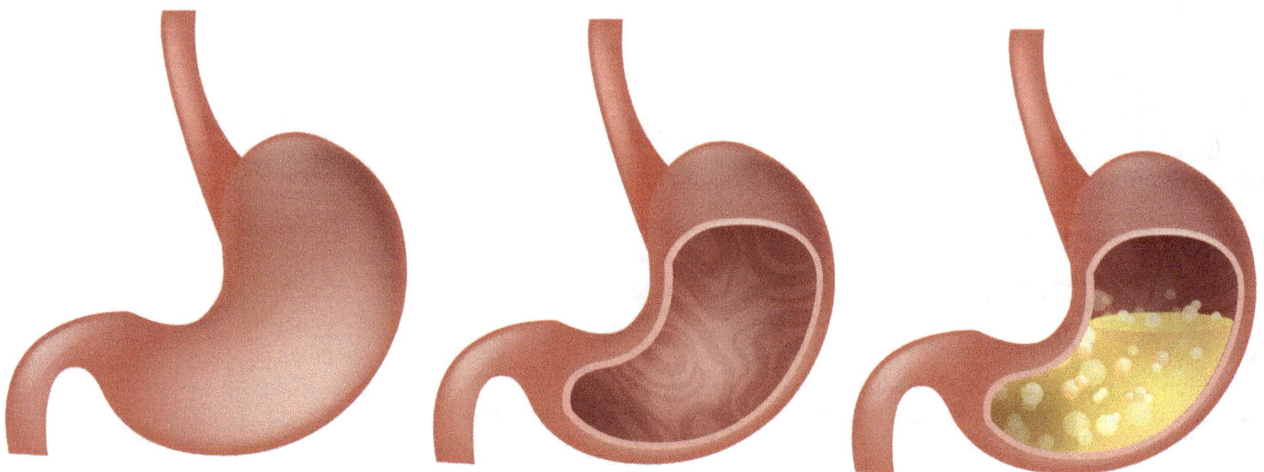

Questions: Stomach

1. How long does it take for the stomach to empty its contents into the small intestine?
2. What two substances are used to change food into liquid?
3. What happened to the bread in the plastic bag?

Stomach
Homework

Directions

Each of these foods strengthen your stomach.
Circle the word if you eat or drink one of these foods this week.

Yogurt

Kefir

Miso

Sauerkraut

Kimchi

Sourdough bread

Almonds

Olive oil

Kombucha

Peas

Brussel sprouts

Bananas

Garlic

Ginger

Take a small step every day!

Chapter 18
Small and Large Intestines
Digestive System

Do you not understand that everything that goes into the mouth passes into the stomach, and is eliminated?
(Matthew 15:17)

Vocabulary

Small intestine: sends nutrients to the cells via the bloodstream

Large intestine: dehydrates unused food and removes it from the body as waste material

Dehydrate: to remove water

Macronutrients (nutrients): carbohydrates, fats, and proteins

Micronutrients (nutrients): minerals and vitamins

Small and Large Intestines

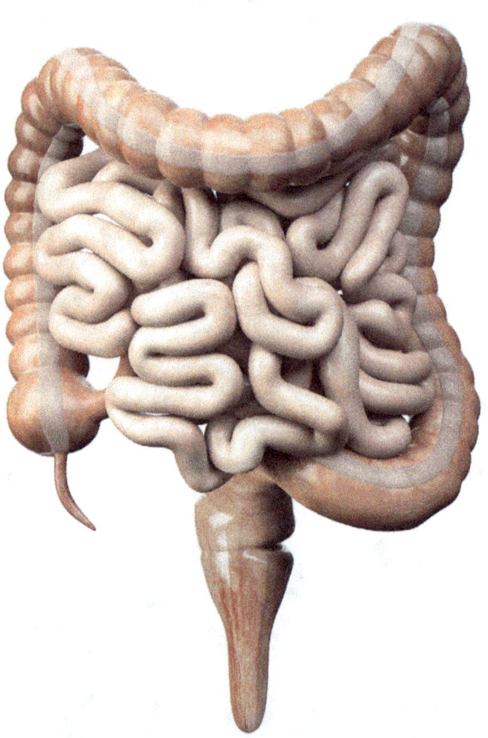

Food (nutrients) goes from the stomach to the small intestine. Food not used by the body goes into the large intestine to be processed into waste material.

Dehydration
Activity

Purpose
To demonstrate the effects of dehydration in the large intestine

Materials
Potato
2 Containers
Cutting board
Knife
2 tablespoons of salt
Water
2 pieces of paper
Pen, pencil, or marker

Procedure
With a permanent marker label "salted" on one of the containers.
Label "plain" on the second container.
Fill each container with an equal amount of water.
Stir two tablespoons of salt into the container marked "salted".
Using a knife and cutting board, cut a potato in half.
Place half of the potato, flat-side down, into the container marked "salted."
Place the other half of the potato, flat side down, into the other container.
Let potatoes sit for one week.
Observe results.

Result
The students will observe a difference in the potato halves. The potato in plain water will be mostly unchanged; the potato in the salty water will be shriveled.

Why?
The salt water acts as a dehydrator by drawing moisture out of the potato. As the water is drawn out of the potato, the potato becomes dehydrated and will shrivel.

The large intestine dehydrates undigested food into waste that is sent out of the body.
However, it needs plenty of water for the process of dehydration to work well. Be sure to drink water!!!

Nutrients

Macronutrients

Carbohydrates
Fats
Proteins

Micronutrients

Minerals and Vitamins

Small Intestine and Large Intestine

Dehydration

Questions: Small and Large Intestines

1. Where does food go after the stomach?
2. What is the main job of the small intestine?
3. Where does food go after the small intestine?
4. What is the main job of the large intestine?
5. What is dehydration?
6. What are the three macronutrients?
7. What are the two micronutrients?

Small and Large Intestines
Homework

Directions

Each of these foods strengthen your small and large intestines.
Circle the word if you eat or drink one of these foods this week.

Small Intestine

Apples

Raspberries

Beans

Peas

Carrots

Celery

Large Intestine

Yogurt

Fruits and Veggies

Salmon

Oatmeal

Chicken

Fish

Cinnamon

Olive Oil

Good food! Good mood!

Chapter 19
Liver, Gallbladder, and Pancreas
Digestive System

And you shall take all the fat that covers the entrails, and the lobe of the liver, and the two kidneys and the fat that is on them, and offer them up in smoke on the altar.
(Exodus 29:13)

Vocabulary

Liver: cleanses blood and helps digestion by secreting bile

Gallbladder: stores bile produced by the liver

Bile: a liquid that breaks down fats

Pancreas: creates hormones to balance blood sugar levels in the blood

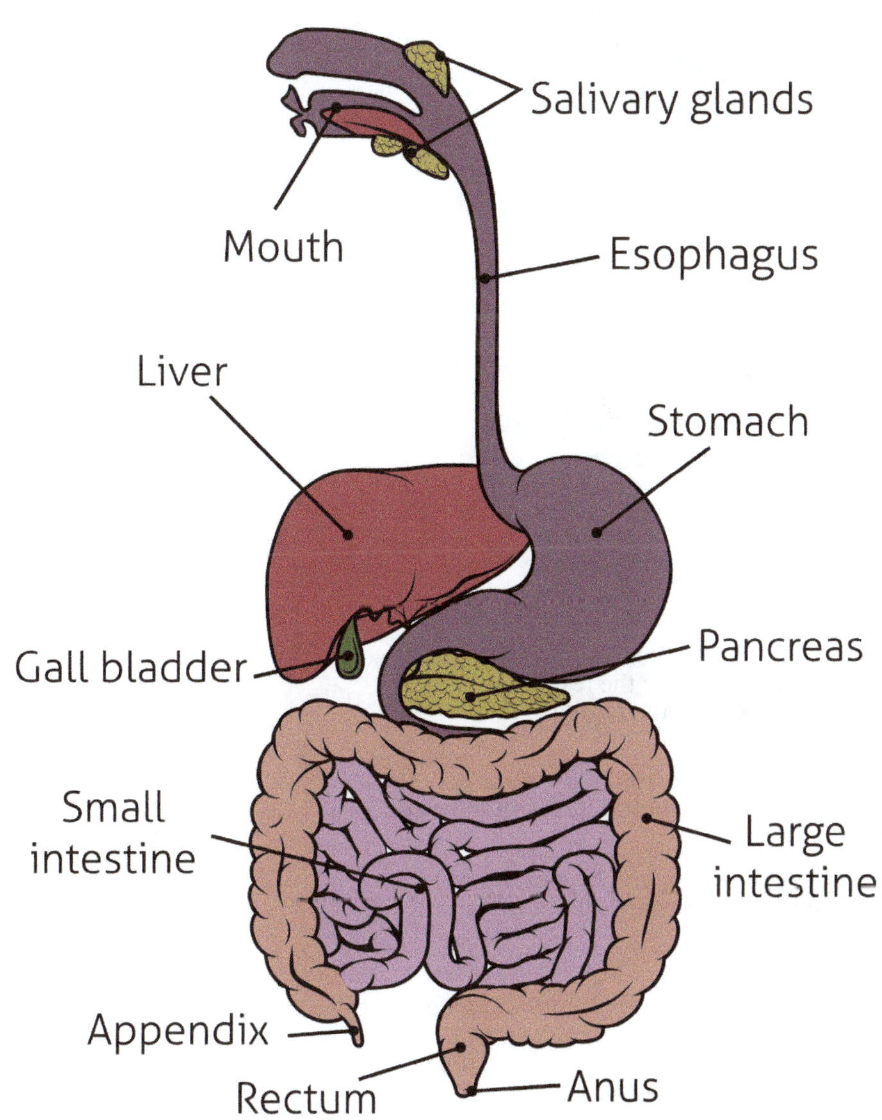

Bile
Activity

Purpose
To show what bile does in the liver and gallbladder

Materials
Jar with a lid
Water
1/4 C Oil
1/4 C Dishwashing soap

Procedure
Fill a jar with water.
Pour oil in the water.
Put a lid on the jar and shake it. Observe the results.
Add soap to jar and shake it. Observe the results.

Result
In the beginning the oil will float to the top of the water.
When soap is added to the water, the oil will mix with the water.

Why?
Soap breaks down the fats in oil through a chemical reaction.
In much the same way, bile breaks down fat in our bodies.

Liver, Gallbladder, and Pancreas
Activity

Unscramble the letters and write the word on the line.

1. REVIL ___________________

2. REDDALBLLAG ___________________

3. SAERCNAP ___________________

4. ELIB ___________________

5. GANOR ___________________

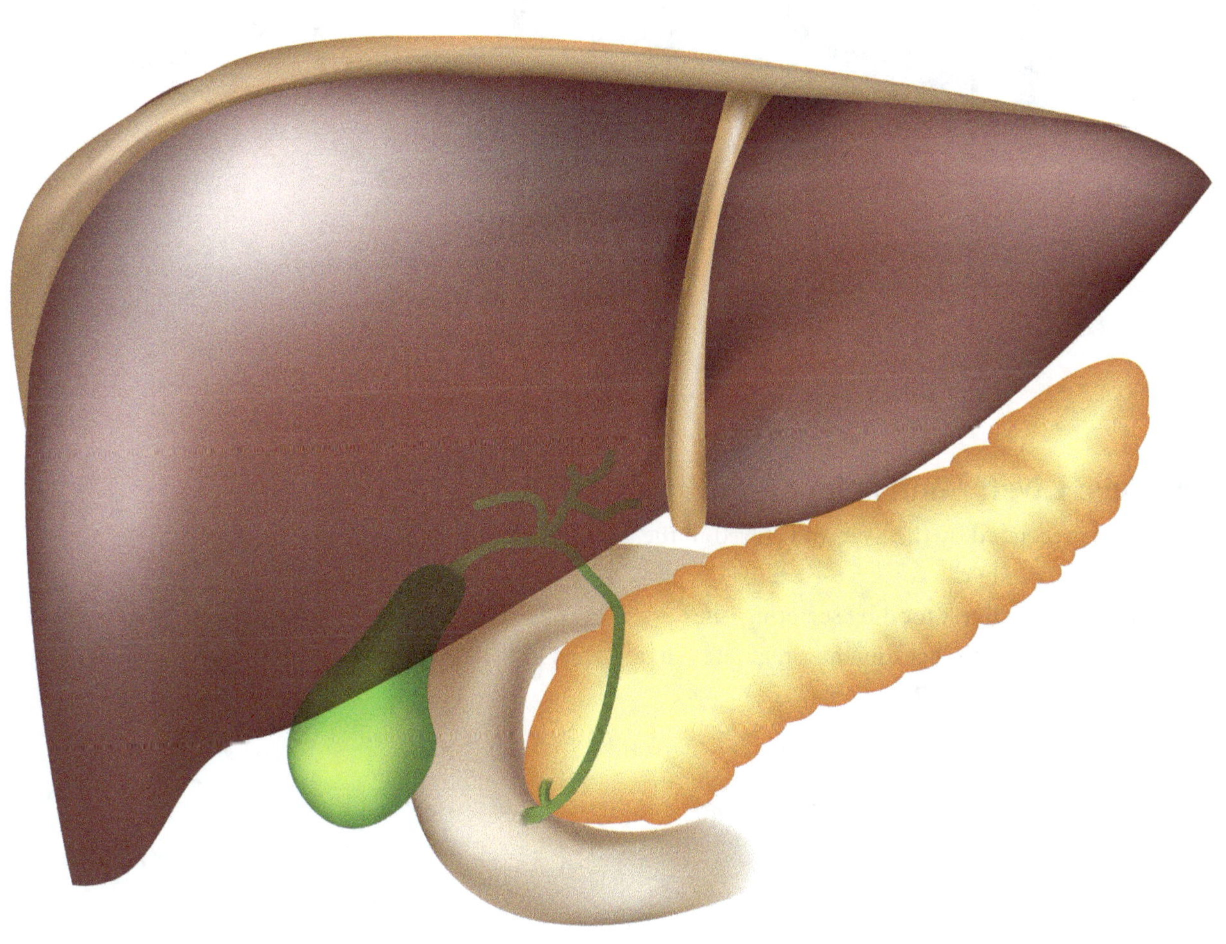

Liver, Gallbladder, and Pancreas

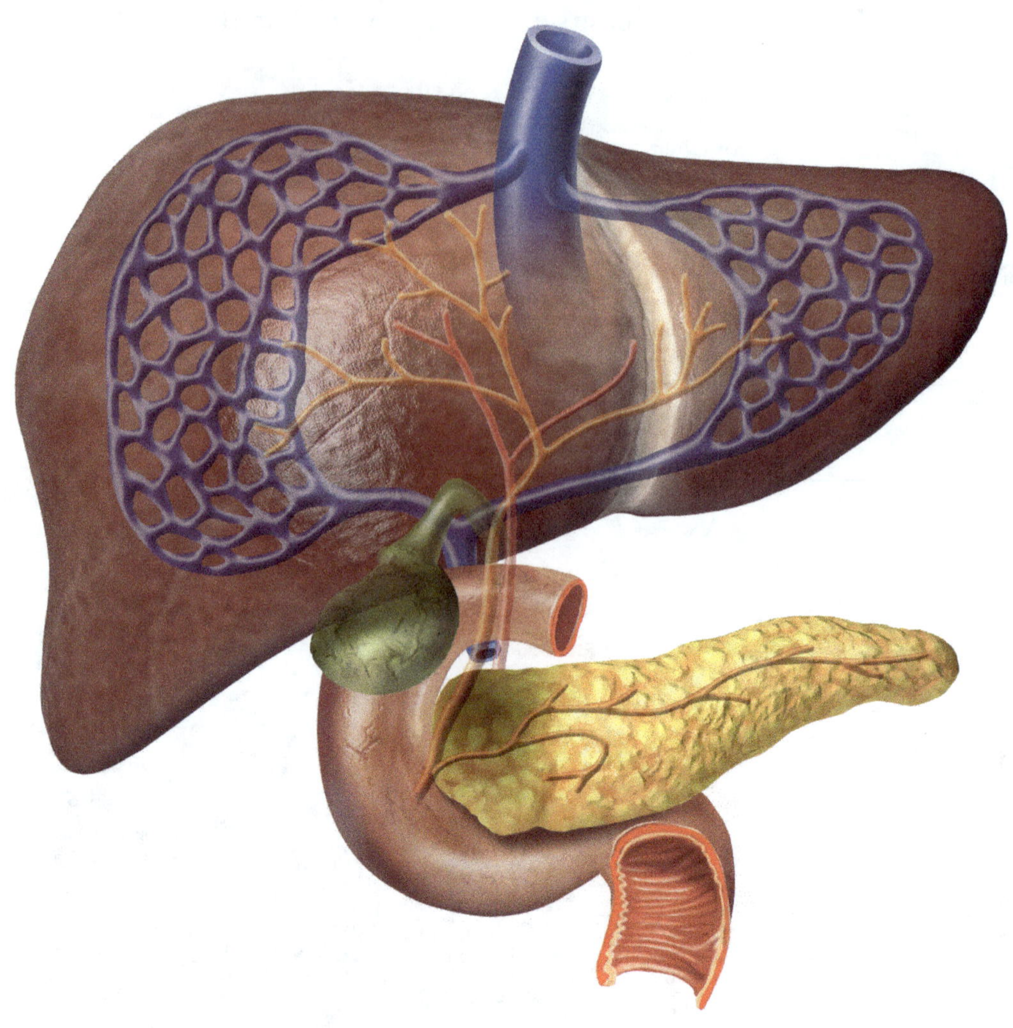

Questions: Liver, Gallbladder, and Pancreas

1. What fluid is secreted by the liver and stored in the gallbladder?
2. What organ stores bile?
3. What organ helps digest food by secreting bile?
4. What organ helps with digestion and regulates sugar levels in the blood?

Chapter 20
Jeopardy Review

Goal: To create **Questions** from **Answers** instead of **Answers** from **Questions**

On a white board, write the categories and point values provided below.

Divide the students into teams.
Use the **Answers** and **Questions** found on pages 99-100.
The teams will take turns until all the **Answers** and **Questions** are used.
Rotate students so everyone gets a turn.
Keep team scores on a white board.

Procedure: A student from each team will choose a category and a point amount.
The teacher will read the **Answer** connected to this choice.
This student will try to think of a **Question** that goes with the **Answer.**

If the student comes up with the correct **Question,**
he or she will receive full point value.
If the student cannot come up with the **Question,**
then ALL teams will concur with each other in separate parts of the room.

ALL the teams with the correct **Question** connected to the **Answer,** will get half the point value.
The **Question** can be written or whispered into the teacher's ear.

The game continues until all the categories and point values
have been chosen and the correct **Questions** have been provided.
The team with the most points wins.

Mouth	Stomach	Intestines	Liver, Gallbladder, and Pancreas
10	10	10	10
20	20	20	20
30	30	30	30
40	40	40	40
50	50	50	50

Jeopardy Questions
Digestive System

Mouth and Esophagus

10 Points
Answer: The starting place of digestion where food enters the body
Question: What is the mouth?

20 Points
Answer: Small white structures covered with enamel
Question: What are teeth?

30 Points
Answer: The tube connecting the mouth to the stomach
Question: What is the esophagus?

40 Points
Answer: The fluid the salivary glands produce
Question: What is saliva?

50 Points
Answer: The substance that covers the outside of teeth
Question: What is enamel?

Stomach

10 Points
Answer: The place food goes after the stomach
Question: What is the small intestine?

20 Points
Answer: Colorless fluids used by the stomach to break down food
Question: What are acids?

30 Points
Answer: Proteins used by the stomach to break down food
Question: What are enzymes?

40 Points
Answer: The organ that mixes, liquifies, and moves food where it needs to go
Question: What is your stomach?

50 Points
Answer: The amount of time it takes for the stomach to digest food
Question: What is 40 minutes to 2 hours?

Large and Small Intestines

10 Points
Answer: The word that means to remove water
Question: What is dehydrate?

20 Points
Answer: The two micronutrients
Question: What are minerals and vitamins?

30 Points
Answer: The three macronutrients
Question: What are carbohydrates, fats, and proteins?

40 Points
Answer: Dehydrates unused food and removes it from the body as waste material
Question: What is the large intestine?

50 Points
Answer: Sends nutrients to the cells via the bloodstream
Question: What is the small intestine?

Liver, Gallbladder, and Pancreas

10 Points
Answer: The liver, gallbladder, and pancreas are in this system.
Question: What is the digestive system?

20 Points
Answer: Digests food by secreting bile
Question: What is the liver?

30 Points
Answer: The organ that stores bile
Question: What is the gallbladder?

40 Points
Answer: The fluid is secreted by the liver and stored in the gallbladder
Question: What is bile?

50 Points
Answer: Helps with digestions and regulates sugar levels
Question: What is the pancreas?

Chapter 21
Immune System

Do you not know that those who run in a race all run, but only one receives the prize?
Run in such a way that you may win.
(1 Corinthians 9:24)

Vocabulary

Immune System: a system used to fight diseases and infections

Spleen: filters blood and stores white blood cells, red blood cells, and platelets

Thymus gland: produces hormones to help with the body's defense against disease

Bone marrow: the place where red blood cells, white blood cells, and platelets are created

Red blood cells: transports oxygen from the lungs to the cells and carbon dioxide from the cells to the lungs

White blood cells: cells created inside bone marrow to help fight disease

Platelets: cells that help wounds heal by forming blood clots to slow or stop bleeding

BLOOD CELL

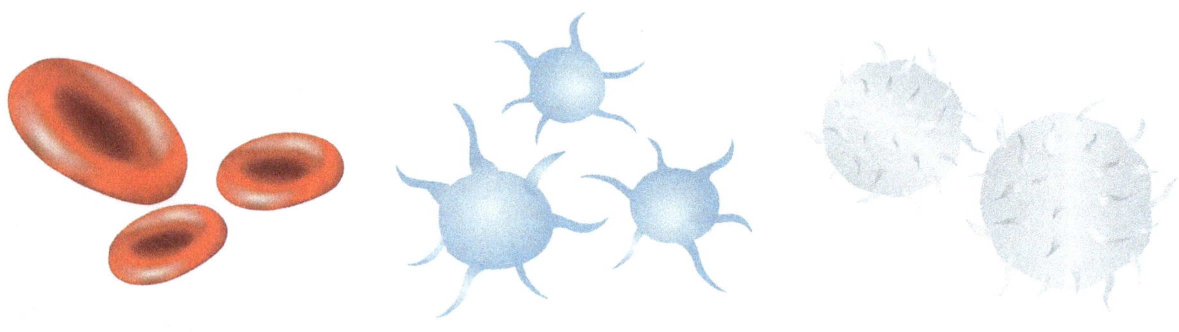

Red blood cells | Platelets | White blood cells

Artificial Germs
Activity

Purpose
To understand how germs spread

Materials
Glo Germ Gel (fake germs)
Glo Germ Powder (fake germs)
UV Flashlight

Procedure
Rub Glo Germ Gel into the palms of your hands; shake your students' hands.
Turn out the light and shine the UV light onto their hands.

Rub Glo Germ Gel on the students' hands (about the size of a dime).
Shine the UV light in a dark room onto their hands.

Wash hands with soap and water.

Shine UV light on the students' hands again to discover how well they cleaned them.

Wash hands as many times as necessary to get rid of "germs."

Sprinkle Glo Germ Powder onto a table.

Turn off the lights and shine the UV light onto the surface.
Clean the table with soap and water until there are no more signs of "germs."
Check the door knobs and walls with UV light looking for "germs" that may have spread.

Result
The gel will cause hands, tables, and other surfaces to "light up"
even after washing with soap and water, especially in fingernails and chapped skin.

Why?
In order to prevent the spread of disease, this activity is used to train medical workers in good hygiene practices.

Immune System

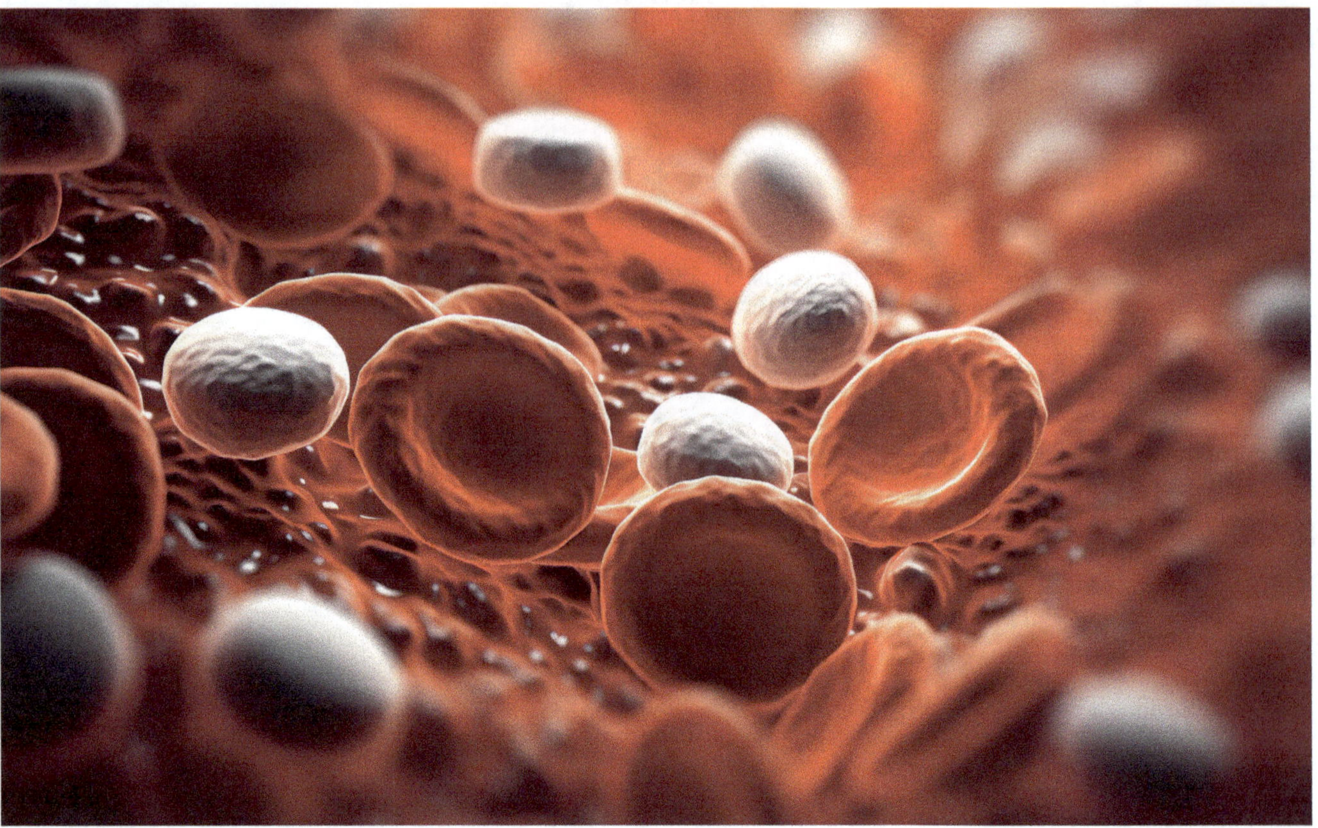

Questions: Immune System

1. Where are red and white blood cells created?
2. What system fights infections and diseases?
3. What helps wounds heal by forming blood clots?
4. What kind of blood cells carry oxygen?
5. What kind of blood cells fight infections and diseases?
6. What organ filters blood and stores white blood cells?
7. What gland releases hormones to help with the body's defense against disease

Immune System
Homework

Directions

Each of these foods strengthen your immune system.
Circle the word if you eat or drink one of these foods this week..

Yogurt
Oats
Barley
Chicken Soup
Fish
Beef
Sweet Potatoes
Kale
Bell Peppers
Eggs
Mushrooms
Kefir

Let food be your medicine!

Chapter 22
Integumentary System

Charm is deceitful and beauty is vain,
But a woman who fears the LORD, she shall be praised.
(Proverbs 31:30)

Vocabulary

Integumentary System: a system used to protect from injury with skin, nails, and hair

Skin: the largest organ in the body; a protective barrier

Nails: a thin hard layer covering the outer tips of fingers and toes

Hair: fine threads located all over a body

Facts

- Skin is the largest organ.
- Skin protects the body from infection and other dangers.
- Skin regulates temperature.
- Skin helps the sense of touch.
- Skin needs protection from the sun (examples: sunscreen, hats, shade).
- Skin heals when damaged.
- Mouth wounds heal the quickest.

SKIN
First Aid

Clothe me with skin and flesh…
(Job 10:11a)

Purpose
To gain the ability to care for "pretend" wounds

Materials
Soap
Water
Clean cloth
Antibiotic cream
Bandage

Procedure
Wash your hands with soap to avoid infection.
Press a cloth on a pretend wound to stop the bleeding.
Rinse a "wound" under running tap water.
Apply an antibiotic cream.
Cover the "wound" with a bandage.

Result
Hands-on first-aid training will be demonstrated and practiced.

Why?
Platelets clot the blood to stop bleeding, white blood cells remove dead and injured cells, and new healthy cells repair the damaged tissue.

Best practice:
Try not to touch another person's blood unless absolutely necessary.

Skin

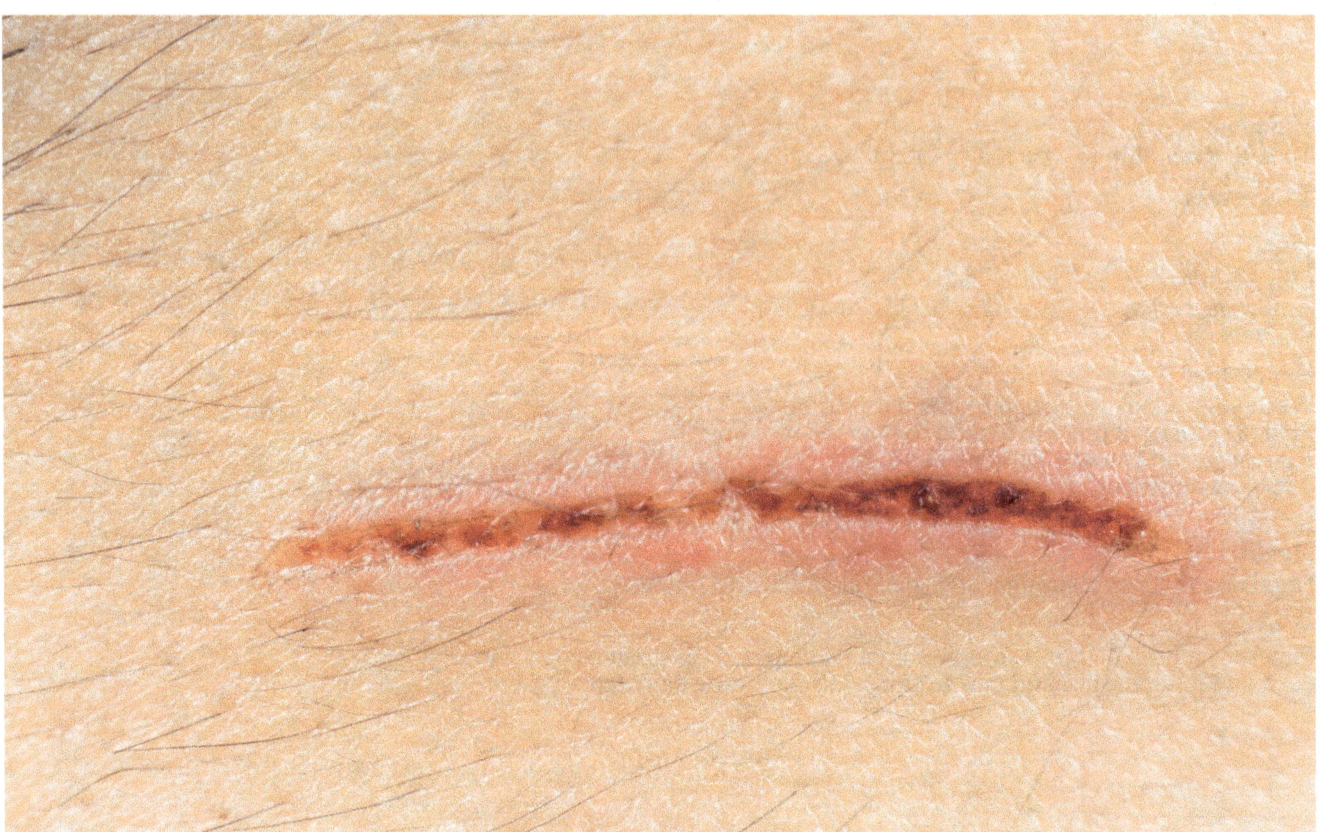

Questions: Integumentary System

1. What is the largest organ in the body?
2. What are two jobs the skin does?
3. What would happen if the body did not have skin?
4. How can skin be kept safe from the sun?
5. What happens when skin is cut?
6. How should a cut be treated?
7. What happens to a cut after a few days?
8. What part of the body heals the quickest?

Integumentary System
Homework

Directions

Each of these foods strengthen your immune system.
Circle the word if you eat or drink one of these foods this week.

Salmon

Watermelon

Green tea

Tomatoes

Carrots

Avocados

Olive oil

Walnuts

Oranges

Kale

Almonds

Eggs

Milk

Peppers

Berries

Broccoli

Sardines

Dark chocolate

Greek yogurt

Eat good!
Feel good!

Chapter 23
Endocrine System

*Do you not know that in a race all the runners run, but only one receives the prize?
So run that you may obtain it. Every athlete exercises self-control in all things.
They do it to receive a perishable wreath, but we are imperishable. So, I do not run aimlessly;
I do not box as one beating the air. But I discipline my body and keep it under control,
lest after preaching to others I myself should be disqualified.
(1 Corinthians 9:24-27)*

Vocabulary

Endocrine system: a system used to make the hormones that help cells talk to each other

Hormones: chemicals produced by the body

Hypothalamus gland: controls the endocrine system; regulates temperature, hunger, thirst, and more

Pituitary gland: receives directions from the hypothalamus; produces hormones that regulate vital body functions

Pineal gland: produces melatonin to help with sleep

Thyroid gland: releases hormones to help regulate growth and metabolism

Parathyroid gland: controls the amount of calcium in blood and bones

Thymus gland: produces hormones to help with the body's defense system against disease

Adrenal gland: regulates responses to stress

Pancreas: creates hormones to balance blood sugar levels in blood

Insulin: a hormone that balances sugar levels in the blood

Hormones Plus Exercise

Purpose
To raise awareness of exercise in relationship to hormones

Materials
Just you and a place big enough to exercise (outdoors or gym)

Procedure
Run laps.
Play freeze tag.
Jog up and down stairs with variations of sprints.
Sprint back and forth pretending to be different animals.

Result
Exercise leads to a healthier lifestyle. It makes the body stronger, the mind sharper, and the mood lighter.

Why?
Regular physical activity stimulates hormones throughout the body.
This helps the body work well and gives a sense of well-being.

ENDOCRINE SYSTEM

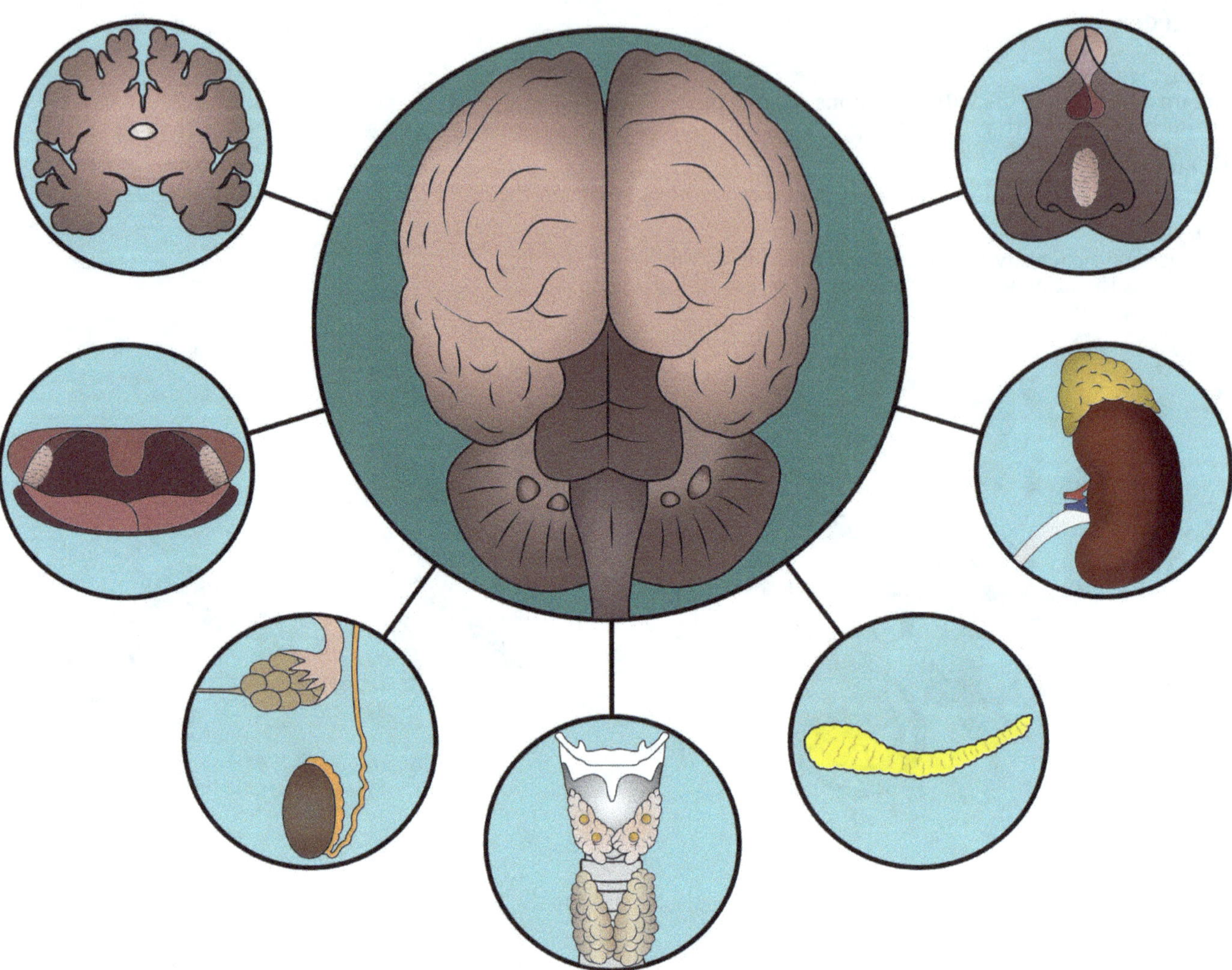

Questions: Endocrine System

1. What gland regulates responses to stress?
2. What hormone helps the body use sugar for energy?
3. What organ produces insulin?
4. What gland releases growth hormones?
5. What gland releases hormones to regulate metabolism?
6. What gland helps with sleep?
7. What gland releases hormones to help defend against diseases?
8. What does the hypothalamus gland do?

Endocrine System
Homework

Directions

Each of these foods strengthen your endocrine system.
Circle the word if you eat or drink one of these foods this week.

Meat

Cauliflower

Brussels sprouts

Bok Choy

Radishes

Kale

Olive oil

Avocado oil

Nut butter

Raw unsalted nuts or seeds

Avocados

Spinach

Kefir

Fermented vegetables

Olives

Laughter is the best medicine!

Chapter 24
Reproductive System

Have you not read that He who created them from the beginning made them male and female.
(Matthew 19:4)

Vocabulary

Reproductive System: a system used to create hormones and new life

Chromosomes: located inside the nucleus of a cell; contains genes

Genes: the basic units of heredity passed from generation to generation

Karyotype: an individual's complete set of chromosomes

Male: XY chromosomes

Female: XX chromosomes

Human karyotype

Male

Female

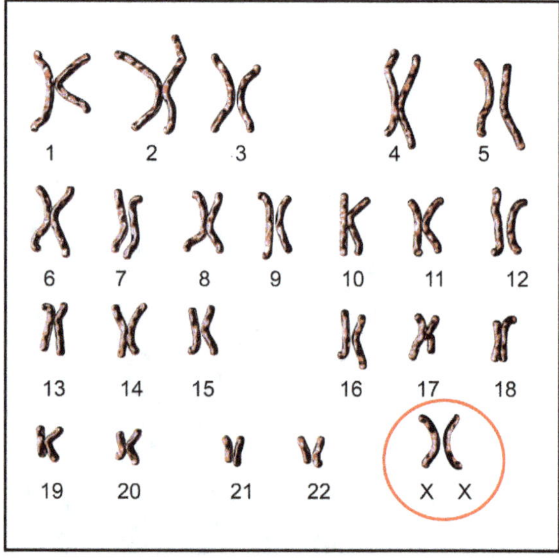

Every cell in your body determines whether you are male or female. This cannot be changed. Females have two X chromosomes; males have one X and one Y chromosome.

Genetics
Create a Monster
Activity

Purpose
To demonstrate how traits are passed down to children by their mothers and fathers

Materials
Dry erase marker, colored pencils, markers, or crayons
White computer paper
Coin

Procedure
Draw a line making two columns on a white board.
Write FATHER at the top of the first column and MOTHER at the top of the second column.
Write different genetic traits under FATHER and MOTHER.

For example:
MOTHER: female, three eyes, curly hair, big nose, big feet, fat, red eyes, eight legs, fangs, and green hair etc.

FATHER: male, one eye, straight hair, small nose, small feet, purple eyes, two legs, crooked teeth, and red hair etc.

Create a monster by flipping a coin: heads/father's trait; tails/mother's trait
Circle the traits after each coin toss.
Create and color monsters using the circled words.

Result
The students will obtain the beginning knowledge of genetics and heredity.

Why?
Genetics are the building blocks of life determining our appearance.

Bingo Card
Activity

Purpose
To develop an understanding of genetics

Materials
Bingo cards
Coins or beans
Genetic traits (cut-outs)
Basket

Procedure
Print and cut out genetic traits; place in a basket.
Print Bingo cards.
Give each student a Bingo card. (They can be the same.)
Choose a trait from the basket and read it.
Place a coin (or bean) on any of the traits that apply.
The first student with four coins or beans in a row wins.

Result
Students will obtain a beginning knowledge of genetics.

Why?
Genetics are the building blocks of life and determine our appearance.

Generation of Green Eyes

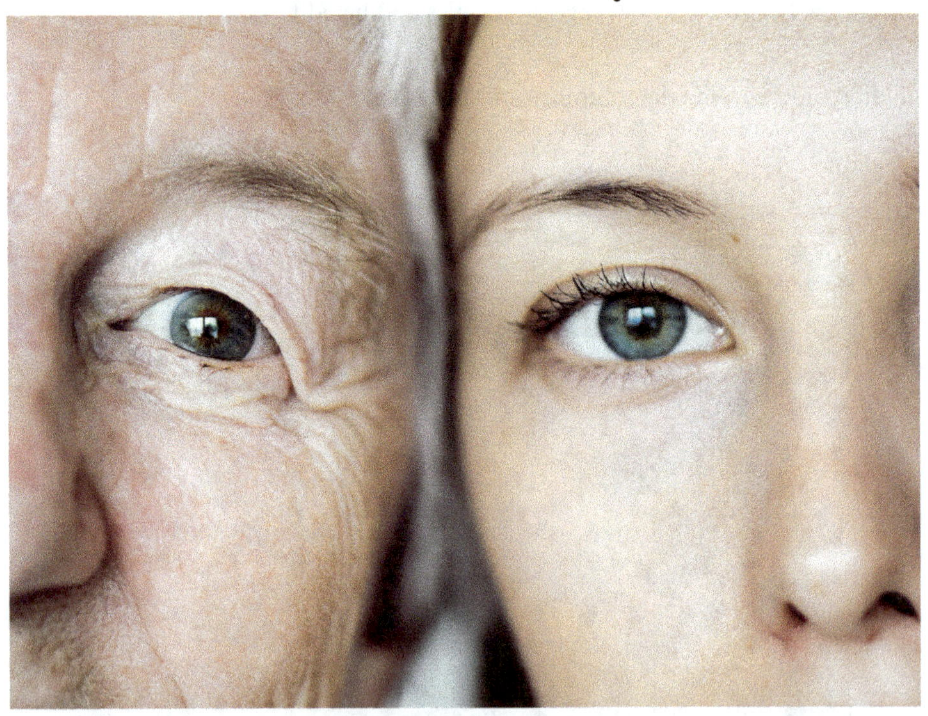

BINGO
Activity

Cut and place in a basket.

FREE SPACE	BROWN HAIR	HAND CLASPING LEFT THUMB ON TOP	FREE SPACE
CURLY HAIR	HAND CLASPING RIGHT THUMB ON TOP	RED HAIR	BROWN EYES
BLUE EYES	LIGHT SKIN	MEDIUM BROWN SKIN	M SHAPED HAIRLINE
STRAIGHT HAIR	DARK SKIN	FREE SPACE	DIMPLES
FRECKLES	BLOND HAIR	SHORT	BLACK HAIR
STRAIGHT ACROSS HAIRLINE	LEFT-HANDED	TALL	GREEN EYES
FREE SPACE	"TACO" TONGUE	RIGHT-HANDED	FREE SPACE

BINGO CARDS

CURLY HAIR	LEFT-HANDED	TALL	FREE SPACE
FREE SPACE	GREEN EYES	MEDIUM BROWN SKIN	RIGHT-HANDED
RED HAIR	STRAIGHT HAIR	BLACK HAIR	SHORT
M SHAPED HAIRLINE	"TACO" TONGUE	FREE SPACE	DARK SKIN
BLOND HAIR	HAND CLASPING LEFT THUMB ON TOP	BLUE EYES	HAND CLASPING RIGHT THUMB ON TOP
STRAIGHT ACROSS HAIRLINE	BROWN HAIR	LIGHT SKIN	FREE SPACE
FREE SPACE	BROWN EYES	DIMPLES	FRECKLES

FREE SPACE	LIGHT SKIN	DIMPLES	RIGHT-HANDED
LEFT-HANDED	TALL	HAND CLASPING LEFT THUMB ON TOP	GREEN EYES
CURLY HAIR	SHORT	FREE SPACE	DARK SKIN
HAND CLASPING RIGHT THUMB ON TOP	FREE SPACE	BLOND HAIR	"TACO" TONGUE
BLACK HAIR	BROWN HAIR	BROWN EYES	M SHAPED HAIRLINE
FREE SPACE	RED HAIR	STRAIGHT ACROSS HAIRLINE	STRAIGHT HAIR
FRECKLES	BLUE EYES	MEDIUM BROWN SKIN	FREE SPACE

BLUE EYES	STRAIGHT HAIR	MEDIUM BROWN SKIN	FREE SPACE
FREE SPACE	BLACK HAIR	HAND CLASPING CLASPING LEFT THUMB ON TOP	"TACO" TONGUE
CURLY HAIR	LEFT-HANDED	TALL	RIGHT-HANDED
FRECKLES	STRAIGHT ACROSS HAIRLINE	BROWN EYES	RED HAIR
GREEN EYES	FREE SPACE	SHORT	M SHAPED HAIRLINE
DIMPLES	DARK SKIN	BLOND HAIR	FREE SPACE
FREE SPACE	HAND CLASPING RIGHT THUMB ON TOP	BROWN HAIR	LIGHT SKIN

GREEN EYES	HAND CLASPING LEFT THUMB ON TOP	STRAIGHT HAIR	FREE SPACE
FREE SPACE	BLUE EYES	RED HAIR	TALL
SHORT	BROWN EYES	BLOND HAIR	M SHAPED HAIRLINE
DIMPLES	FREE SPACE	BROWN HAIR	LIGHT SKIN
HAND CLASPING RIGHT THUMB ON TOP	BLACK HAIR	RIGHT-HANDED	LEFT-HANDED
FREE SPACE	FRECKLES	"TACO" TONGUE	MEDIUM BROWN SKIN
STRAIGHT ACROSS HAIRLINE	CURLY HAIR	DARK SKIN	FREE SPACE

Color Blind Test

Can you see all the numbers in the circles?

If you are color blind, you see colors differently than most people.
It usually means you have a hard time telling the difference between certain colors, but not all colors.
Color blindness is usually an inherited trait passed from mother to son.
However, girls can sometimes be color blind.

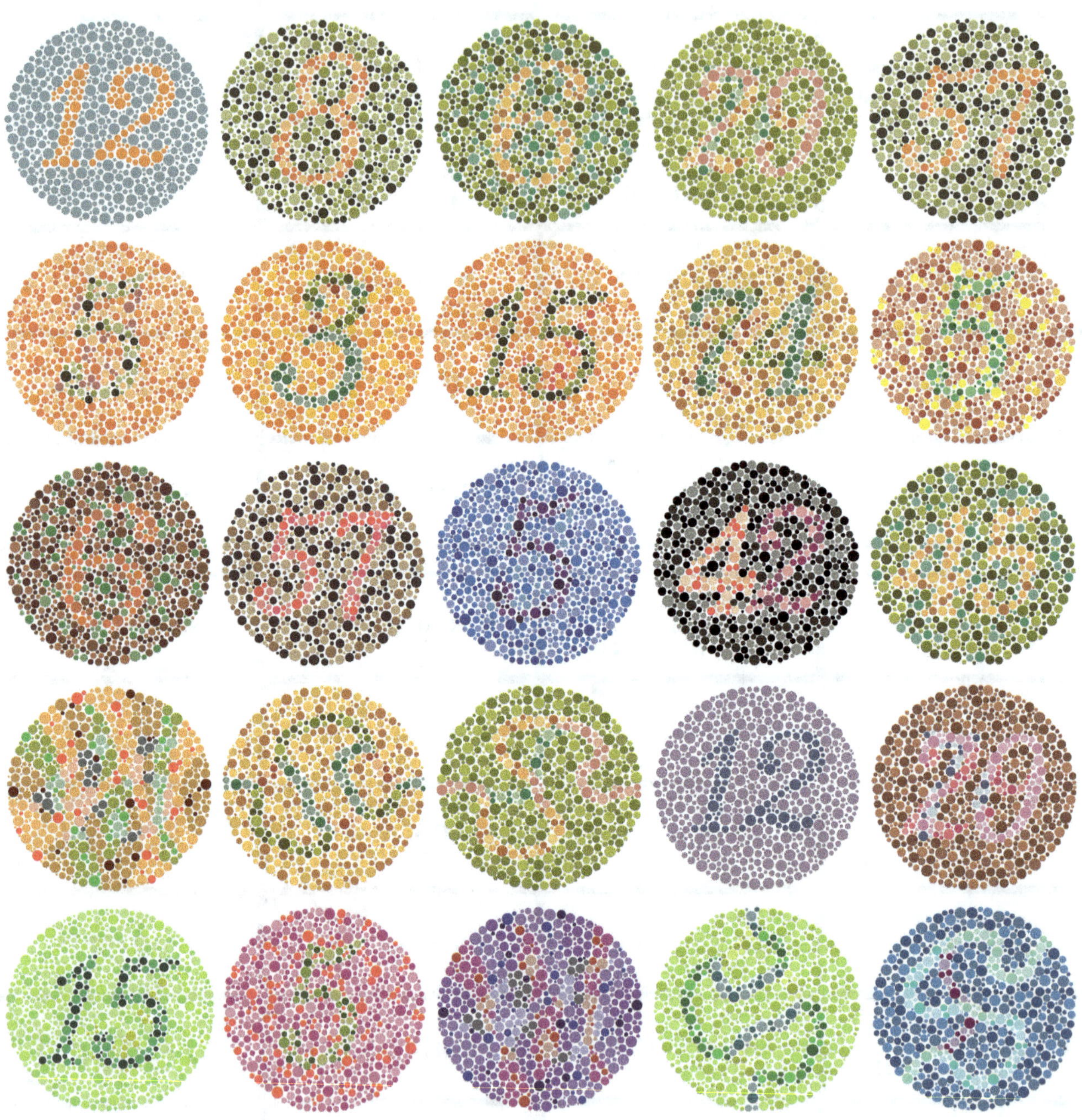

The Development of a Baby

First Trimester

First three months of pregnancy

Month 1

An amniotic sac gradually fills with fluid and forms around a fertilized egg.
The placenta begins to transfer food (nutrients) from the mother to her baby.
The baby's mouth, lower jaw, and throat begin to develop.
The circulation system begins.
By the fourth week, the baby is smaller than a grain of rice.

Month 2

Facial features start to develop. Ears, eyes, fingers, toes, arms, and legs form.
The nervous system, digestive system, and sensory organs develop.
The heartbeat can be heard around six weeks.

Month 3

At nine weeks all the major body systems are formed.
The arms, hands, fingers, feet and toes are fully developed. Outer ears are visible.
Circulatory and urinary systems begin working.
By the end of the first trimester, the baby is approximately 4 inches and weighs about 1 ounce.

Second Trimester

Second three months of pregnancy

Month 4

Facial features develop more fully.
An ultrasound can show whether the baby is a boy or a girl.
The heartbeat can be heard by a doppler.
Fingers and toes can be seen.
Eyelids, eyebrows, eyelashes, nails, and hair are formed.
Teeth and bones become stronger.

Month 5

The mother begins to feel her baby move.
The baby develops muscles and exercises them.
Hair begins to grow.

Month 6

The baby responds to sounds.
The baby can live outside the womb if born this early.
By the end of the second trimester, the baby is about 12 inches long and weighs about 2 pounds.

Month 7

The baby matures and develops reserves of body fat.
The baby responds to sound, pain, and light.

Third Trimester

Last three months of pregnancy
A full-term pregnancy is about 40 weeks.

Month 8

The baby continues to mature and gain weight.
Most internal organs are developed. The lungs may still be immature.

Month 9

Lungs are fully developed.
The baby can blink, close eyes, grasp firmly, and respond to sounds, light, and touch.
By thirty-six weeks, the baby weighs between 5 ½ to 6 ½ pounds.

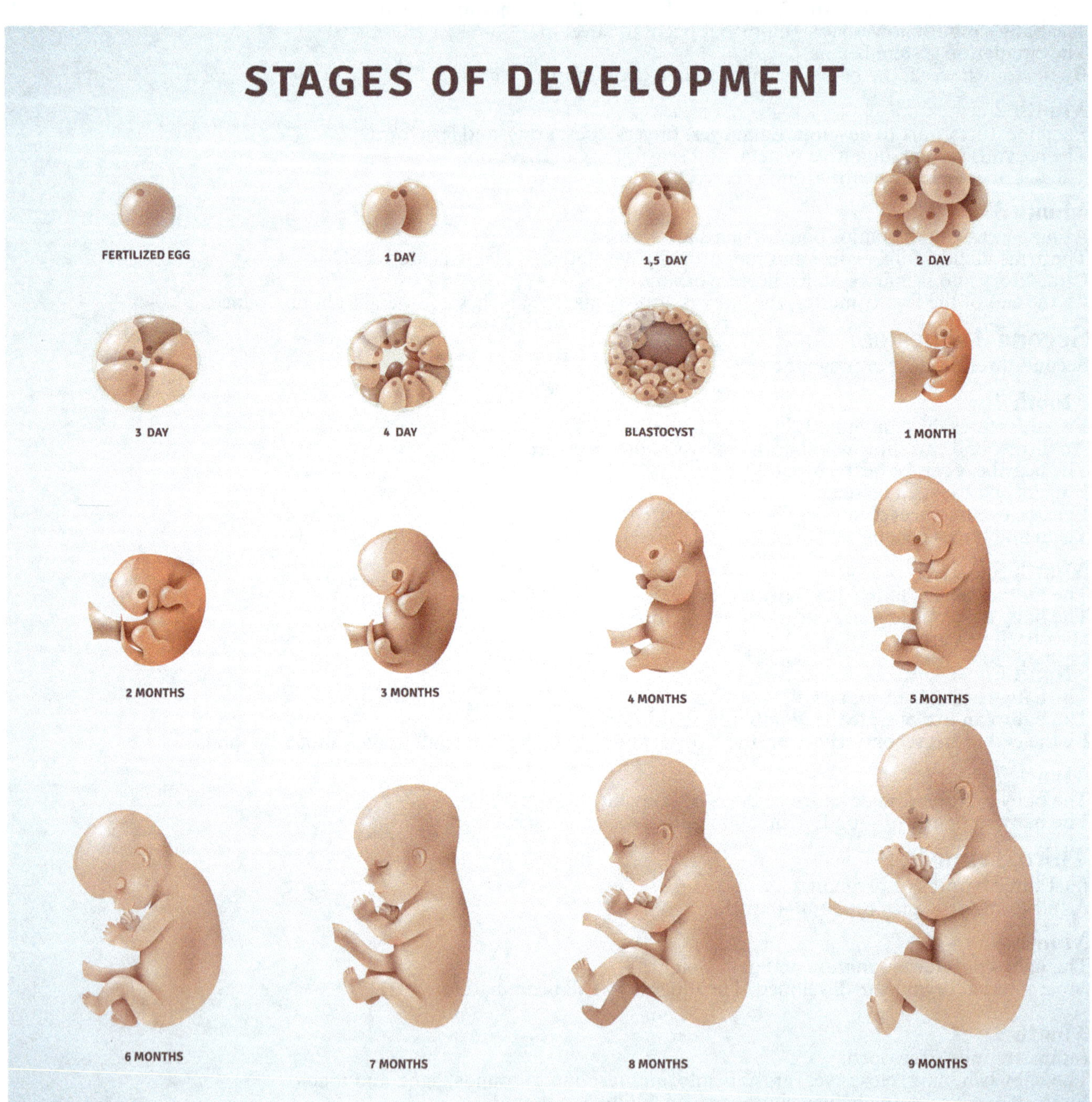
STAGES OF DEVELOPMENT
FERTILIZED EGG
1 DAY
1,5 DAY
2 DAY
3 DAY
4 DAY
BLASTOCYST
1 MONTH
2 MONTHS
3 MONTHS
4 MONTHS
5 MONTHS
6 MONTHS
7 MONTHS
8 MONTHS
9 MONTHS

Genetics

Can you curl your tongue?

The ability to make a "taco" tongue is genetic.

Questions: Reproductive System

1. How many trimesters (stages) are there in a woman's pregnancy?
2. What month do facial features start to develop?
3. What month does a baby respond to sound, pain, and light?
4. What month does the mother begin to feel the baby move?
5. What month are the baby's lungs fully developed?
6. Can you make a "taco" tongue? Can your parents make a "taco" tongue?

Chapter 25
Jeopardy Review

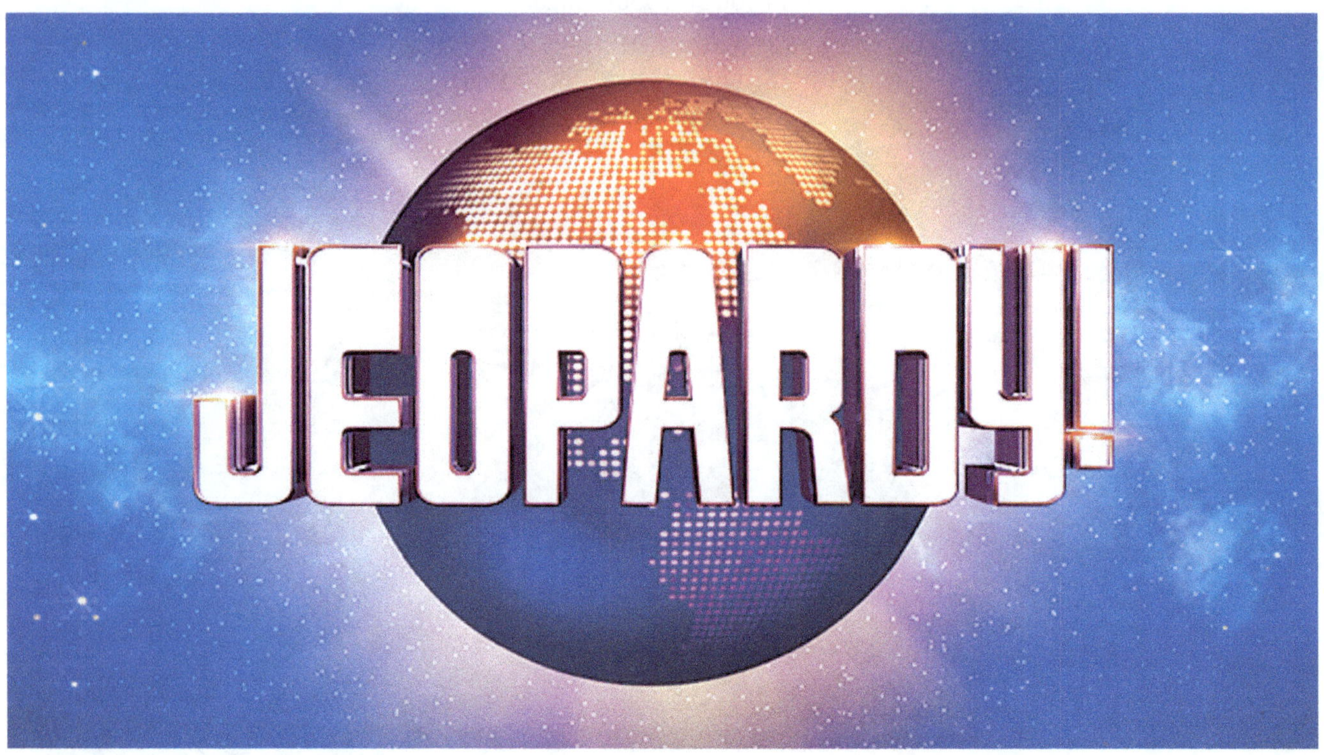

Goal: To create **Questions** from **Answers** instead of **Answers** from **Questions**

On a white board, write the categories and point values provided below.

Divide the students into teams.
Use the **Answers** and **Questions** found on pages 124-125.
The teams will take turns until all the **Answers** and **Questions** are used.
Rotate students so everyone gets a turn.
Keep team scores on a white board.

Procedure: A student from each team will choose a category and a point amount.
The teacher will read the **Answer** connected to this choice.
This student will try to think of a **Question** that goes with the **Answer.**

If the student comes up with the correct **Question,**
he or she will receive full point value.
If the student cannot come up with the **Question,**
then ALL teams will concur with each other in separate parts of the room.

ALL the teams with the correct **Question** connected to the **Answer,** will get half the point value.
The **Question** can be written or whispered into the teacher's ear.

The game continues until all the categories and point values
have been chosen and the correct **Questions** have been provided.
The team with the most points wins.

IMMUNE	INTEGUMENTARY	ENDOCRINE	REPRODUCTIVE
10	10	10	10
20	20	20	20
30	30	30	30
40	40	40	40
50	50	50	50

Immune System

10
Answer: The place where red and white blood cells are created
Question: What is bone marrow?

20
Answer: The type of blood cells that carry oxygen
Question: What are red blood cells?

30
Answer: The type of blood cells that fight disease
Question: What are white blood cells?

40
Answer: The organ that filters and stores white blood cells
Question: What is the spleen?

50
Answer: Helps wounds heal by forming blood clots to slow or stop bleeding
Question: What are platelets?

Integumentary System

10
Answer: A method to keep safe from the sun
Question: What is sunscreen, hats, or shade? (Answers may vary)

20
Answer: The largest organ in the body
Question: What is skin?

30
Answer: The part of the body that heals the quickest
Question: What is the mouth?

40
Answer: Fine threads located all over the body
Question: What is hair?

50
Answer: A hard covering at the tips of fingers
Question: What are nails?

Endocrine System

10
Answer: Regulates stress responses
Question: What is the adrenal gland?

20
Answer: Chemicals produced to help regulate the body
Question: What are hormones?

30
Answer: The hormone needed to use sugar for energy
Question: What is insulin?

40
Answer: Creates insulin to regulate blood sugar levels in the blood
Question: What is the pancreas?

50
Answer: Releases hormones to help regulate growth and metabolism
Question: What is the thyroid gland?

Reproductive

10:
Answer: The number of trimesters in a pregnancy
Question: What is three?

20:
Answer: A person who sees colors differently
Question: What is colorblind?

30:
Answer: Another word for inherited traits such as eye color
Question: What are genes?

40:
Answer: An individual's complete set of chromosomes
Question: What is a karyotype?

50:
Answer: Located inside the nucleus of a cell; contains genes
Question: What are chromosomes?

My Awesome Body

Teacher's Helps

Science with Grandma

Supplies
Answers
Vocabulary
Images and Sources

My Awesome Body

Table of Contents

Exploring With Grandma Publishing Company, Virginia

exploringwithgrandma@gmail.com

exploringwithgrandma.com

Supplies

Body Systems Overview
Song: *Twelve Body Systems* (Hopscotch: YouTube)
Vocabulary: Twelve Body Systems (page 5)
Model or picture of a car
Matching: Twelve Body Systems (page 7)
Word Search: Twelve Body Systems (page 8)
Word Search: Make Your Own (page 9)

Chapter 1: Skeletal System
Labeled: Skeletal System (page 12)
Unlabeled: Skeletal System (page 13)
Questions: Skeletal System (page 15)
Food homework: Skeletal System (page 16)

Chapter 2: Muscular System
Word Search: Muscular System (page 20)
Questions: Muscular System (page 21)
Food homework: Muscular System (page 22)

Chapter 3: Nervous System Overview
Questions: Nervous System (page 24)
Questions: Helen Keller (page 26)
Food homework: Nervous system (page 27)

Chapter 4: Sense of Sight - Nervous and Muscular Systems
Activity: Drawing a face with your eyes closed (page 29)
Optical illusions (page 30)
Colored pencils, markers, or crayons
Questions: The Eye (page 32)
Homework: Spiritual application on the eye (page 33)

Chapter 5: Sense of Sound - Nervous System
Multiple jars
Food coloring (different colors)
Water
Spoon
Questions: The Ear (page 36)
Homework: Sounds (page 37)

Chapter 6: Sense of Smell - Nervous System
Smell chart: Essential oils (page 39)
Questions: Sense of Smell (page 40)
Homework: Sense of Smell (page 41)

Chapter 7: Sense of Touch - Nervous System
Three plastic tubs
Water (hot, room temperature, and cold)
Ice
Questions: Sense of Touch (page 44)
Homework: Sense of Touch (page 45)

Chapter 8: Sense of Taste - Nervous System
Taste chart: Tastes (page 47)
Questions: Sense of Taste (page 48)
Homework: Sense of Taste (page 49)

Chapter 9: Jeopardy Review
Questions: Skeletal, Muscular, and Nervous Systems (page 51)

Chapter 10: Respiratory System
White cardboard
2 copies of the lungs and face (page 54)
Black marker
2 balloons
Two plastic straws
Tape
Glue
Questions: Respiratory System (page 55)
Food homework: Respiratory System (page 56)

Chapter 11: Circulatory System

Colored labeled heart: (page 58)
Labeled heart (page 59)
Unlabeled heart (page 60)
Pulse chart (page 62)
Questions: Circulatory System (page 63)
Food homework: Circulatory System (page 64)

Chapter 12: Lymphatic System

Unscramble and fill in the blank: Lymphatic System (page 67)
Questions: Lymphatic System (page 68)

Chapter 13: Urinary System

pH scale (pages 70-71)
pH strips
saliva (spit)
Questions: Urinary System
Water and food homework: Urinary System (page 73)

Chapter 14: Jeopardy Review

Questions: Circulatory, Lymphatic, and Urinary Systems (pages 75-76)

Chapter 15: Digestive System Overview

Word Search: Digestive System (page 78)
Questions: Digestive System (page 79)
Food homework: Digestive System (page 80)

Chapter 16: Mouth and Esophagus - Digestive System

5 clear glasses
5 eggs
Water
Apple juice
Cola
Lemon juice
Vinegar
Toothpaste
Toothbrush
Questions: Mouth and Esophagus (page 83)
Food homework: Mouth and Esophagus (page 84)

Chapter 17: Stomach - Digestive System

White bread
1/2 cup water
1/2 cup apple cider vinegar
Small plastic bags
Questions: Stomach (page 87)
Food homework: Stomach (page 88)

Chapter 18: Small and Large Intestines - Digestive System
Potato
2 Containers
Cutting board
Knife
2 tablespoons of salt
Water
2 pieces of paper
Pen, pencil, or marker
Questions: Small and Large Intestines (page 92)
Food homework: Small and Large Intestines (page 93)

Chapter 19: Liver, Gallbladder, and Pancreas - Digestive System
Jar with a lid
Water
1/4 C Oil
1/4 C Dishwashing soap
Unscramble: Liver, Gallbladder, and Pancreas (page 96)
Questions: Liver, Gallbladder, and Pancreas (page 97)

Chapter 20: Jeopardy Review
Questions: Mouth, Stomach, Intestines, Liver, Gallbladder, and Pancreas (pages 99-100)

Chapter 21: Immune System
Glo Germ Gel (fake germs)
Glo Germ Powder (fake germs)
UV Flashlight
Questions: Immune System (page 103)
Food homework: Immune System (page 104)

Chapter 22: Integumentary System
Soap
Water
Clean cloth
Antibiotic cream
Bandage
Questions: Integumentary System (page 107)
Food homework: Integumentary System (page 108)

Chapter 23: Endocrine System
Questions: Endocrine System (page 111)
Food homework: Endocrine System (page 112)

Chapter 24: Reproductive System

Dry erase marker, colored pencils, markers, or crayons
White computer paper
Coin
Genetic traits (cut-outs) (page 116)
Bingo cards (pages 117-118)
Coins or beans
Basket
Colorblind Test (page 119)
Questions: Reproductive System (page 122)

Chapter 25: Jeopardy Review

Questions: Immune, Integumentary, and Reproductive Systems

Answers

Overview (page 7)

1. D
2. C
3. L
4. K
5. E
6. I
7. J
8. B
9. H
10. F
11. G

Chapter 1: Skeletal System (page 15)

1. ribs
2. pleasant words
3. skeletal system
4. white blood cells
5. vertebrates
6. bend, twist, stretch, and stand-up straight
7. both have seven bones
8. bone
9. the place where two bones come together in order to move
10. ligaments
11. protects brain, eyes, ears, and the rest of your face
12. baby: 300 bones; adult: 206 bones
13. red blood cells
14. platelets

Chapter 2: Muscular System (page 21)

1. skeletal, smooth, and cardiac
2. heart
3. digestive tract and blood vessels
4. biceps and triceps
5. blood vessels, digestive tract, and heart
6. heart
7. 640
8. tendons connect muscles to bones;
ligaments connect bones to other bones
9. healthy food and exercise

Chapter 3: Nervous System Overview (pages 24; 26)

1. nervous system
2. brain
3. neurons (nerve) cells
4. spinal cord
5. see, hear, smell, touch, and taste

1. 19 months
2. 7 years
3. Anne Sullivan
4. Water pump

Chapter 4: Sense of Sight (page 32)

1. nervous and muscular systems
2. cornea
3. optical illusion
4. far-sighted
5. near-sighted
6. pupil
7. retina

Chapter 5: Sense of Sound (page 36)

1. outer ear
2. ear canal
3. vibrations
4. cochlea
5. ear drum
6. auditory nerve

Chapter 6: Sense of Smell (page 40)
1. nervous system
2. nasal cavity
3. nostrils
4. olfactory receptors
5. sense of smell
6. up to 10,000
7. children
8. African elephant
9. Covid 19

Chapter 7: Sense of Touch (page 44)
1. nervous system
2. dendrites
3. axons
4. nucleus
5. myelin sheath

Chapter 8: Sense of Taste (page 48)
1. sour, sweet, bitter, salty, and savory
2. taste buds
3. bitter
4. sour
5. savory
6. sweet
7. salty
8. Covid 19

Chapter 10: Respiratory System (page 55)
1. exhale
2. inhale
3. No, they are two different tubes. The esophagus is used for food. The trachea (windpipe) is used for air.
4. oxygen
5. carbon dioxide
6. lungs

Chapter 11: Circulatory System (page 63)
1. circulatory system
2. heart
3. away from the heart
4. aorta
5. bright red
6. back to the heart
7. vena cavas - superior and inferior
8. dark red
9. about 60,000 miles; 2.5 times around the world
10. capillaries
11. lungs

Chapter 12: Lymphatic System (page 67-68)
1. tonsil
2. marrow
3. blood
4. spleen
5. lymph
6. cells
7. fluid
8. blood
9. bone marrow
10. exercise

1. adnoids and tonsils
2. fights infections with white blood cells; manages fluid levels
3. spleen
4. lymph nodes
5. thymus gland

Chapter 13: Urinary System (page 72)
1. bladder
2. kidneys
3. ureter
4. urethra

Chapter 15: Digestive System Overview (page 79)
1. mouth
2. saliva
3. esophagus
4. stomach
5. small intestine
6. large intestine
7. to remove water
8. the liver
9. the gallbladder
10. breaks down fats

Chapter 16: Mouth and Esophagus (page 83)
1. mouth
2. teeth, saliva, and tongue
3. saliva
4. tongue
5. esophagus
6. enamel
7. brushing, drinking water, and using teeth only for chewing

Chapter 17: Stomach (page 87)
1. between 40 minutes and 2 hours
2. acids and enzymes
3. turns to liquid

Chapter 18: Small and Large Intestines (page 92)
1. small intestine
2. sends nutrients to the cells via the bloodstream
3. large intestine
4. dehydrates undigested food into waste
5. water removed from food
6. carbohydrates, fats, and proteins
7. minerals and vitamins

Chapter 19: Liver, Gallbladder, and Pancreas (pages 96-97)
1. LIVER
2. GALLBLADDER
3. PANCREAS
4. BILE
5. ORGAN

1. bile
2. gallbladder
3. liver
4. pancreas

Chapter 21: Immune System (page 103)
1. bone marrow
2. immune system
3. platelets
4. red blood cells
5. white blood cells
6. spleen
7. thymus

Chapter 22: Integumentary System (page 107)
1. skin
2. answers may vary: protects bodies from danger, regulates temperature, allows our sense of touch, heals when damaged etc.
3. no protection; death
4. answers may vary: sun screen, hats, clothes, and umbrellas
5. Platelets clot our blood to stop bleeding; white blood cells remove the dead and injured cells; new healthy cells repair the damaged tissue.
6. wash, press, rinse, apply antibiotics, and cover
7. heals
8. mouth

Chapter 23: Endocrine System (page 111)
1. adrenalin
2. insulin
3. pancreas
4. pituitary gland
5. thyroid gland
6. pineal gland
7. thymus gland
8. controls temperature, hunger, and thirst

Chapter 24: Reproductive System (page 122)
1. three
2. month two
3. month seven
4. month five
5. month nine
6. answers vary

TWELVE BODY SYSTEMS
WORD SEARCH
ANSWERS

A	I	N	T	E	G	U	M	E	N	T	A	R	Y
S	D	F	E	N	I	R	C	O	D	N	E	G	R
H	Y	R	E	S	P	I	R	A	T	O	R	Y	E
J	R	K	L	W	E	N	T	R	U	I	O	Z	P
X	O	R	C	V	B	A	E	N	M	W	E	R	R
E	T	Y	O	T	U	R	I	R	C	O	L	P	O
V	A	F	G	T	H	Y	J	I	V	A	K	L	D
I	L	D	E	S	A	A	T	Z	T	O	X	C	U
T	U	R	N	E	W	A	M	E	N	B	U	V	C
S	C	T	U	Y	H	U	L	I	O	P	L	S	T
E	R	Z	M	P	A	E	S	C	D	G	H	J	I
G	I	X	M	C	K	V	B	N	R	M	O	P	V
I	C	Y	I	S	W	A	S	C	G	I	V	B	E
D	L	R	R	A	L	U	C	S	U	M	C	N	M

Page 8

URINARY	IMMUNE	INTEGUMENTARY
RESPIRATORY	DIGESTIVE	MUSCULAR
NERVOUS	CIRCULATORY	REPRODUCTIVE
LYMPHATIC	ENDOCRINE	SKELETAL

MUSCLES
WORD SEARCH
ANSWERS

A	V	S	M	O	O	T	H	B	C	D	E	F	G
S	Q	Y	S	K	E	L	E	T	A	L	G	D	A
T	R	S	P	O	M	N	L	B	I	C	E	P	S
U	V	T	W	X	Y	U	S	K	N	J	I	H	B
S	K	E	L	Z	A	P	S	B	V	T	C	D	C
W	C	M	V	B	E	M	L	C	O	R	P	A	Y
S	M	A	O	C	N	L	T	K	L	A	U	B	R
F	A	K	I	B	K	I	R	P	U	E	W	C	A
I	B	R	T	D	J	Y	Y	Z	N	H	X	D	T
S	T	M	O	I	R	R	W	E	T	M	A	E	N
I	U	K	G	U	L	A	N	R	A	B	D	F	U
O	S	E	N	O	B	W	C	H	R	S	P	S	L
P	X	C	V	B	N	A	M	V	Y	H	O	L	O
J	A	N	O	D	N	E	T	B	N	M	N	M	V

Page 20

MUSCLE	SYSTEM	SKELETAL
HEART	SMOOTH	BICEPS
VOLUNTARY	CARDIAC	TRICEPS
TENDON	INVOLUNTARY	BONES

DIGESTIVE SYSTEM
WORD SEARCH
ANSWERS

B	F	P	Q	S	E	N	I	T	S	E	T	N	I
A	P	P	E	N	D	I	X	J	R	T	A	B	P
C	D	E	S	Z	V	I	H	O	A	S	D	A	F
G	S	E	W	A	G	H	J	K	L	Z	R	X	C
R	T	H	H	X	L	I	V	E	R	C	E	B	N
S	O	S	Q	Y	M	I	Q	W	R	E	D	R	E
Y	M	U	P	I	D	Y	V	E	U	I	D	E	V
T	A	G	Y	Z	P	R	A	A	Z	X	A	C	I
U	C	A	S	V	N	S	A	B	R	N	L	T	T
H	H	H	A	S	K	D	E	T	F	Y	B	U	S
T	G	P	H	J	K	S	L	A	E	S	L	M	E
U	F	O	G	L	M	U	H	J	K	U	L	D	G
O	W	S	N	E	R	N	T	Y	U	I	A	C	I
M	O	E	P	A	N	C	R	E	A	S	G	W	D

Page 78

GALLBLADDER	DEHYDRATE	ESOPHAGUS
APPENDIX	MOUTH	PANCREAS
INTESTINES	STOMACH	RECTUM
SALIVARY	LIVER	DIGESTIVE

Vocabulary

Acidic: a pH of 0-6

Acids: watery, colorless fluid used by the stomach to break down food

Adenoids: trap bacteria and viruses that enter the mouth and nose

Adrenal gland: regulates responses to stress

Alkaline: a pH of 8 or higher

Aorta: the largest artery in the body

Arteries: blood vessels that take blood away from the heart

Auditory nerve: the nerve that carries an electrical signal to the brain from the ear

Axon: the part of a nerve that takes information away from the cell body

Bile: a liquid that breaks down fats

Bitter: harsh and disagreeable tastes (examples: unsweetened chocolate, coffee, and aspirin)

Bladder: an organ that stores urine

Blood: a liquid that carries nutrients and oxygen throughout the body

Bone: the hard white tissue that makes up the skeleton

Bone marrow: the place where red blood cells, white blood cells, and platelets are created

Brain: an organ inside the head that controls all of the functions in the body

Bronchial tubes: the airways used to let air in and out of the lungs

Capillaries: the smallest blood vessels that connect arteries to veins

Carbon dioxide: an odorless, colorless gas the body considers as waste

Cardiac Muscle: the heart; an involuntary muscle

Cell body: the central part of a nerve cell

Chromosomes: located inside the nucleus of a cell; contains genes

Cochlea: a structure in the inner ear that resembles a snail shell

Cilia: tiny hairs inside the cochlea that respond to different sounds

Cilia: tiny hairs that clean mucus from nasal cavities

Circulatory system: a system used to deliver nutrients and oxygen to all the cells in the body

Cornea: the protective outer layer of the eyes

Dehydrate: to remove water

Dendrite: the part of a nerve that brings electrical signals to the cell body

Digestive System: a system used to break down food for energy, growth, and tissue repair

Diaphragm: a strong muscle under the lungs

Ear: the part of the body used for hearing and balance

Ear canal: a passage made of bone and skin leading to the eardrum

Ear drum: the part of the middle ear that vibrates in response to sound waves

Enamel: a hard, white substance that covers the surface of teeth

Endocrine system: a system used to make the hormones that help cells talk to each other

Enzymes: a type of protein used by the stomach to break down food

Esophagus: a tube used to carry food and liquid from the throat to the stomach

Exhale: to breathe out air

Eye: sensory organs that allow you to see

Far sighted: distant objects are seen clearly, but nearby objects are blurry

Female: XX chromosomes

Gallbladder: stores bile produced by the liver

Genes: the basic units of heredity passed from generation to generation

Hair: fine threads located all over a body

Heart (cardiac): the muscle responsible for pumping blood throughout the body

Hormones: chemicals produced by the body

Hypothalamus gland: controls the endocrine system; regulates temperature, hunger, thirst, and more

Immune System: a system used to fight diseases and infections

Inhale: to breathe in air

Insulin: a hormone that balances sugar levels in the blood

Integumentary System: a system used to protect from injury with skin, nails, and hair

Involuntary muscles: the type of muscles that work without thinking (examples: digestive tract, lungs, and heart)

Joint: the place where two bones come together to move
a. hinge joint - moves back and forth like a door hinge (examples: elbow and knee)
b. ball and socket joint - moves like a joystick (examples: hip and shoulder)

Karyotype: an individual's complete set of chromosomes

Kidneys: filters waste out of the blood and gets rid of it as urine

Large intestine: dehydrates unused food and removes it from the body as waste material

Ligaments: bands connecting bones to other bones

Liver: cleanses blood and helps digestion by secreting bile

Lungs: the organs the body uses to breathe air

Lymph: fluid in the lymphatic system

Lymphatic System: a system used to fight infections with white blood cells; manages fluid levels

Lymphatic vessels: carries lymph and white blood cells

Lymph nodes: bean-shaped glands along lymphatic vessels that filter lymph and fight infections

Macronutrients (nutrients): carbohydrates, fats, and proteins

Male: XY chromosomes

Micronutrients (nutrients): minerals and vitamins

Mouth: the beginning of your digestive system where food enters the body

Mucus: nasal fluid used to trap unwanted substances

Muscles: the parts of the body that work with the skeletal system to contract and move; also provide stability

Muscular system: a system used to give strength, posture, and movement by working with the skeletal system

Myelin sheath: the fatty material that surrounds axons

Nails: a thin hard layer covering the outer tips of fingers and toes

Nasal cavity: the space inside the nose responsible for the movement of air

Near sighted: nearby objects are seen clearly, but far-away objects are blurry

Nervous System: a system used to carry messages between the brain and the body

Neurons (nerves): the cells used to see, hear, smell, touch, and taste

Neutral: a pH of 7

Nose: the part of the body used for breathing and smell

Nostrils: two holes in the nose used for breathing

Nucleus: the control center of a cell

Nutrients: fats, proteins, carbohydrates, vitamins, and minerals

Olfactory nerves: used to send messages to the brain in order to determine smells

Optical illusion: when eyes are tricked into seeing something that is not true

Outer ear: the part of the ear that can be seen

Oxygen: an odorless, colorless gas the body considers as waste

Pancreas: creates hormones to regulate blood sugar levels in blood

Parathyroid gland: controls the amount of calcium in blood and bones

Pineal gland: produces melatonin to help with sleep

Pituitary gland: receives directions from the hypothalamus; produces hormones that regulate vital body functions

Platelets: cells that help wounds heal by forming blood clots to slow or stop bleeding

Pulse: the thump in the arteries whenever the heart beats

Pupil: the black openings in the center of an eye used to let in light

Red blood cells: transports oxygen from the lungs to the cells and carbon dioxide from the cells to the lungs

Reproductive System: a system used to create hormones and new life

Respiratory system: a system used to bring oxygen into the bloodstream and cells

Retina: the part of an eye that senses light and sends images to the brain

Saliva: a clear liquid in the mouth that helps begin digestion; also known as spit

Salivary glands: produces saliva to break down food

Salty: saline (examples: sea water, chips, and pretzels)

Savory: a spicy or salty taste without sweetness (examples: meat, cheese, and pizza)

Skeletal muscle: a type of voluntary muscles attached to bones allowing movement

Skeletal system: a system used to provide structure and give protection to body parts

Skin: the largest organ in the body; a protective barrier

Skull: bones fused together to protect the brain, eyes, ears, and face

Small intestine: sends nutrients to the cells via the bloodstream

Smell: the sense that detects odors using the nose

Smooth Muscle: a type of involuntary muscles (examples: digestive tract and blood vessels)

Spleen: filters blood and stores white blood cells, red blood cells, and platelets

Spinal cord: the nerves, protected by the spine, used to send information to and from the brain

Saliva: a clear liquid; also known as spit

Salivary glands: produces saliva

Skin: a protective covering for the body

Stomach: mixes, liquifies, and breaks down food using acids and enzymes

Sour: having an acidic taste (examples: lemons, limes, and vinegar)

Sweet: food with a high sugar content (examples: cake, honey, and maple syrup)

Taste buds: sensory organs on the tongue used for tasting

Teeth: small, white structures in the mouth used for chewing

Tendons: used to connect muscles to bones

Thymus gland: produces hormones to help with the body's defense system against disease

Thyroid gland: releases hormones to help regulate growth and metabolism

Tongue: the part of your mouth used for tasting, licking, swallowing, and speech

Tonsils: traps bacteria and viruses that enter the mouth and nose

Trachea (windpipe): connects the throat to the lungs

Ureters: the tubes that carry urine from the kidneys to the bladder

Urethra: the tube that rids the body of urine

Urinary System: a system used to filter waste and remove excess fluid from the blood in the form of urine

Veins: the blood vessels that take blood back to the heart

Vena cava: large veins (inferior and superior) that carry blood back into the heart

Vertebrate: protects neurons (nerves) used to help the body bend, twist, stretch, and stand

Vibrate: to move back and forth in order to create sound

Voluntary muscles: the muscles that work by thinking (examples: biceps and triceps)

White blood cells: cells created inside bone marrow to help fight disease

Images and Sources

Images
Exercising Grandma: Jeanette Summerville, 1
Body parts: Jeanette Summerville, 3; 5

Overview
Images
Car: Adobe, 6
Three men: Shutterstock, 7

Sources
Psalm 139:14-17: New American Standard Bible (NASB), 5
I Corinthians 12:12: New American Standard Bible (NASB), 6

Chapter 1: Skeletal System
Images
Pieces of skeleton: Shutterstock, 10
Labeled skeleton: Jeanette Summerville, 11-12
Unlabeled skeleton: Jeanette Summerville, 13
Giraffe: Shutterstock, 14
Spine: Adobe, 14
Skeleton: Adobe, 15
Healthy food: Pixabay, 16

Sources
Proverbs 16:24: New American Standard Bible (NASB), 10
Genesis 2:19-23: New American Standard Bible (NASB), 10
Skeletal system food list: University of Utah, 16

Chapter 2: Muscular System
Images
Muscle man: Adobe, 17
Biceps, triceps, and quadriceps, Shutterstock, 18
Family of muscles: Adobe, 19
Types of muscles, Shutterstock, 21
Wise Grandma: Jeanette Summerville, 22

Sources
Job 40:15-18: New American Standard Bible (NASB), 17
My Body: Susan Ring, 17-21
Muscular system food list: Ortho Arizona/Food for Healthy Bones and Muscles, 22

Chapter 3: Nervous System Overview
Images
Nervous System: Adobe, 23
Boy/Nervous system: Jeanette Summerville, 24
Helen Keller coin: Shutterstock, 25
Helen Keller photograph: Shutterstock, 26
Wise Grandma: Jeanette Summerville, 27

Sources
Helen Keller: Britannica, 25-26
Nervous system food list: Healthier Steps, 27

Chapter 4: Sense of Sight (Nervous and Muscular Systems)
Images
Boy/Nervous system: Jeanette Summerville, 28
Gift of Sight: Jeanette Summerville, 29
Optical illusions: Adobe, 30-31
Eye: Adobe, 32

Sources
Luke 6:42: New American Standard Bible (NASB), 28
I Timothy 5:8: New American Standard Bible (NASB), 33
Matthew 22:39: New American Standard Bible (NASB), 33

Chapter 5: Sense of Sound (Nervous System)

Images

Boy/Nervous system: Jeanette Summerville, 34
Musical jars: Adobe, 35
Anatomy of an Ear: Adobe, 36
Whispering: Adobe, 36

Sources

Romans 10:17: New American Standard Bible, (NASB), 34

Chapter 6: Sense of Smell (Nervous System)

Images

Boy/Nervous system: Jeanette Summerville, 38
Essential oils: Adobe, 39
African elephant: Pixabay, 40
Anatomy of the nose: Adobe, 40

Sources

Exodus 37:29: New American Standard Bible (NASB), 38
Essential oils: The Essential Life/Total Wellness, 39

Chapter 7: Sense of Touch (Nervous System)

Images

Boy/Nervous system: Jeanette Summerville, 42
Neuron: Adobe, 44

Sources

Mark 10:13a: New American Standard Bible (NASB), 42
Nerves: Neuroscience for Kids, 42-44

Chapter 8: Sense of Taste (Nervous System)

Images

Boy/Nervous system: Jeanette Summerville, 46
Tongue: Pixabay, 48

Sources

Psalm 34:8: New American Standard Bible (NASB), 46

Chapter 9: Jeapardy Review

Images

Game logo, Adobe, 50

Chapter 10: Respiratory System

Images

Boy blowing bubbles: Adobe, 52
Lungs: Adobe, 55

Sources

Genesis 2:7: New International Version (NIV), 52
Respiratory system food list: Healthline, 56

Chapter 11: Circulatory System

Images

Heart: Jeanette Summerville, 57
Labeled heart : Jeanette Summerville, 58-59
Unlabeled heart: Jeanette Summerville, 60
Circulatory journey: Jeanette Summerville, 61
Heart: Shutterstock, 63

Sources

Leviticus 17:11: New American Standard Bible (NASB), 57
Circulatory system food list: Healthline, 64

Chapter 12: Lymphatic System

Images

Lymph system: Adobe, 65-66
Lymphatic system: Adobe, 67
Lymphatic system: Shutterstock, 68

Sources

I Corinthians 6:19-20: New American Standard Bible (NASB), 65
Lymphatic system: American Cancer Institute, 65-68
Lymphatic system: Healthline, 65-68
Lymphatic system: Health Direct, 65-68
Lymphatic system: Live Strong, 65-68

Chapter 13: Urinary System

Images

Human Urinary System: Adobe, 69
Alkaline versus Acidic Foods: Adobe, 70
pH scale: Adobe, 71
Cranberry, Adobe, 72
Sources
Exodus 29:13: New International Version (NIV), 69
pH scale: Healthline, 71-72
Urinary system food list: Health Essentials, 73

Chapter 14: Jeapardy Review

Images

Game logo, Adobe, 74

Chapter 15: Digestive System Overview

Images

Digestive system: Adobe, 77
Girl/stomach: Adobe, 79

Sources

Matthew 15:17: New International Version (NIV), 77
Digestive system food list: Healthline, 80

Chapter 16: Mouth and Esophagus

Images

Child/tongue: Shutterstock, 81
Teeth: Pixabay, 82
Mouth x-ray: Shutterstock, 83
Mouth food list: Eat This, 84

Sources

Proverbs 19:24: New American Standard Bible, (NASB), 81

Chapter 17: Stomach

Images

Stomach: Shutterstock, 85
Stomach class experiment: 86
Stomach: Shutterstock, 87
Stomach food list: Healthline, 88

Sources

Matthew 15:17: New American Standard Bible (NASB), 85

Chapter 18: Small and Large Intestine

Images

Small and large intestines: Pixabay, 89
Dehydrated foods: Adobe, 90
Healthy fats: Shutterstock, 91
Healthy protein: Shutterstock, 91
Healthy carbs: Shutterstock, 91
Minerals: Shutterstock, 91
Vitamins: Shutterstock, 91
Chopped food: Pixabay, 92
Wise Grandma: Jeanette Summerville, 93

Sources

Matthew 15:17: New American Standard Bible (NASB), 89
Food list/Small intestine: Live Strong, 93
Food list/large intestine: Colon Surgeons of Charleston, 93

Chapter 19: Liver, Gallbladder, and Pancreas

Images

Digestive System: Adobe, 94
Bile: Shutterstock, 95
Liver, gallbladder, and pancreas: Shutterstock, 96
Liver, gallbladder, and pancreas: Shutterstock, 97

Sources

Exodus 29:13: New American Standard Bible (NASB), 94

Chapter 20: Jeopardy Review

Images

Game logo, Adobe, 98

Chapter 21: Immune System

Images

Blood cells: Shutterstock, 101
Hand-washing: Shutterstock, 102
Red and white blood cells: Adobe, 103
Wise Grandma: Jeanette Summerville, 104

Sources

I Corinthians 6:19: New International Version (NIV), 101
Immune system food list: Prevention, 104

Chapter 22: Integumentary System

Images
Woman's face: Adobe, 105
First aid: Shutterstock, 106
Wound: Adobe, 107
Sources
Proverbs 31:30: New American Standard Bible (NASB), 105
Job 10:11a: New American Standard Bible (NASB), 106
Skin: Kid's Health in the Classroom, 105-107
Wound healing: TCI Medicine, 105-107
Mouth wounds: Science News, 105-107
Wound care: Mayo Clinic, 108
Integumentary system food list: Eat This Not That, 108

Chapter 23: Endocrine System

Images

Hormones: Shutterstock, 109
Marathon: Jeanette Summerville, 110
Endocrine system: Jeanette Summerville, 111
Endocrine system food list: Eating Well, 112
Wise Grandma: Jeanette Summerville, 112

Sources

1 Corinthians 9:24-27: New American Standard Bible (NASB), 109
Endocrine System: National Cancer Institute, 109-111
Endocrine system and exercise: Physiopedia, 109-111

Chapter 24: Reproductive System

Images

Human genes: Adobe, 113
Monster genetics: Adobe, 114
Green eyes genetics: Shutterstock, 115
Color blind test: Adobe, 119
Stages of development: Adobe, 120-121
Rolled tongue: Adobe, 122

Sources

Matthew 19:4: New American Standard Version, 113
Genetics: National Library of Science, 113-122
Colorblind: Healthline, 119
Baby's Journey: Cleveland Clinic, 120-121

Chapter 25: Jeopardy Review

Images

Game logo, Adobe, 123

www.ingramcontent.com/pod-product-compliance
Lightning Source LLC
LaVergne TN
LVHW061204120826
845149LV00011B/1898

* 9 7 9 8 9 9 4 8 9 3 1 8 0 *